THE SILVER LINING

THE SILVER LINING

The Benefits of Natural Disasters

SETH R. REICE

Princeton University Press
Princeton and Oxford

Published by Princeton University Press, 41 William Street,
Princeton, New Jersey 08540
In the United Kingdom: Princeton University Press,
3 Market Place, Woodstock, Oxfordshire OX20 1SY

Second printing, and first paperback printing, 2003
Paperback ISBN 0-691-11368-8

The Library of Congress has cataloged the cloth edition of this book as follows

Reice, Seth Robert, 1947–
The silver lining : the benefits of
natural disasters / Seth R. Reice.
p. cm.
Includes bibliographical references (p.).
ISBN 0-691-05902-0 (cloth : alk. paper)
1. Ecological disturbances. 2. Natural disasters—
Environmental aspects. I. Title.
QH545.N3 R45 2001
577.2—dc21   2001016376

British Library Cataloging-in-Publication Data is available

This book has been composed in New Baskerville

Printed on acid-free paper. ∞

www.pupress.princeton.edu

Printed in the United States of America

10 9 8 7 6 5 4 3 2

Contents

Preface

All the world's a stage, and all the
men and women merely players.

—William Shakespeare, *As You Like It,*
Act II, Scene 7

At the beginning of the new millennium, we are faced with a seemingly endless series of environmental crises. From global warming to water pollution, most of these problems are of our own doing. Ecology is the study of the interrelationships between organisms and their environment, so all environmental problems are ecological problems. Our ability to address the environmental problems of the new millennium depends on understanding the ecology of the issues involved and our role in them.

At this critical juncture, at a time when we need an ecological perspective most, the field of ecology is changing. For decades we believed that the most important ecological processes were determined by the interactions among organisms. Our stories, and legends, and religions are filled with images of animals inflicting catastrophic change—foxes in the henhouse killing and eating chickens; a plague of locusts devastating wheat crops. The interactions between predators and their prey and between herbivores and plants were part of a conceptualization and understanding of nature that raised biological interactions to preeminent status among all ecological processes. In that view, the environment was merely a backdrop.

In the new worldview, sometimes called the "New Ecology," the environment is seen as an active participant. Even the often violent disturbances in the environment such as fires, floods, hurricanes, tornadoes, landslides, earthquakes, and volcanic eruptions that we call disasters are now being appreciated as critical actors in the entire "ecological play" (as termed by G. Evelyn Hutchinson). Far more chickens are lost every year to extreme heat and drought than to foxes. Crops are far more frequently destroyed by storms than by locusts. Ecologists have come to a new understanding of the role of natural disasters in the dynamics of ecosystems.

Most of us have lived through major "disasters" such as the eruption of Mount Saint Helens twenty-one years ago. At first this appeared to be an unmitigated disaster, devastating an entire landscape and rendering it lifeless. But even there, on the plains of Mount Saint Helens, many diverse species have survived and now thrive. So, was it really a disaster? Or was it just part of the natural dynamics of ecosystems that we now call "disturbance ecology"? This book is aimed at bringing the fresh insights of disturbance ecology to a broader, nontechnical audience. My goal in this book is to develop and clarify the general reader's understanding of natural disturbances and their role in nature. I will offer fresh perspectives from disturbance ecology, which gives us a new way of looking at nature.

Disturbances are paradoxical. What we see and fear is their destructive power, yet these same disturbances help create and maintain the biodiversity that benefits both the ecosystem and ourselves. Diverse, healthy ecosystems provide essential services to humankind. From producing clean air to breathe and clean water to drink, the ecosystems of the Earth make human life possible. Disturbances enhance the capacity of the ecosystems to do these indispensable tasks. If we can incorporate these new ideas about disturbances and disasters into our thinking, we will change the way we act in our world. If we can understand how ecosystems work, then we can effectively manage them.

Shakespeare reminds us that "all the world's a stage, and all the men and women merely players." In nature, the stage is the entire biosphere. Humans are just one of the vast cast of characters in the play, although, as people, we often have leading roles. The "special effects" of storms and lightning are not offstage but are a significant part of the action on the stage. Fires and floods impact not only forests and rivers, but entire landscapes and the people in it.

Optimists claim that every cloud has a silver lining. I go even further. I will show that every tornado's funnel cloud, every forest fire's billowing cloud of smoke, indeed every disturbance, has tremendous benefits to the ecosystem it impacts. This is the real silver lining.

As we attempt to solve environmental problems, we must first understand our roles in nature. By trying to resist the forces of nature (for example, by damming rivers and suppressing forest fires), we have caused many of our problems and made others much worse. If people can understand how ecosystems really work and how vital is the role of natural disasters, then we can learn to live with them. If we learn to live harmoniously with nature, all people will have a brighter future. Showing you the importance of living with uncontrolled natural systems and processes is the ultimate goal of this book.

In working on this project, I have many people to thank. First and foremost, I thank my wife, Sheila Rice Evans, whose unflagging support and encouragement made this book possible. My daughter, Rebecca Rose Reice, has listened patiently as I tried out phrases and sections on her. Jack Repcheck, my first editor at Princeton University Press, suggested the book and encouraged me through the first draft. Kristin Gager, my subsequent editor, did the hard work of transforming a rough manuscript into a book, doing so with an admirable blend of firmness and sensitivity. Finally, Sam Elworthy was instrumental in transforming the final manuscript into a real book. My mom, Sylvie Reice, an editor herself, offered a lot of helpful feedback.

Three ecologists have helped to shape the ideas in this book. Conrad Istock taught me that a scientist has a responsibility to society. Joe Connell taught me that disturbances are important to biodiversity, and that doing experiments is the way to get real answers in ecology. Bill Cooper, my major professor, taught me that it is important to communicate to all people, and taught me how.

Thanks to you all.

THE SILVER LINING

Chapter 1 ***The More Things Change, the More They Stay the Same***

Nature is an endless combination and repetition of a very few laws. She hums the old well-known air through innumerable variations.

—Ralph Waldo Emerson, "History," in *Essays* (First Series, 1841)

Yellowstone National Park, August 20, 1988: "Black Saturday"

It was very hot and dry. Winds had been blowing from the west for weeks. These hot winds sucked the moisture from the leaves, from the trees, from the Earth itself. An ancient 100-foot-tall Lodgepole Pine bent with the force of the wind. Its sap grew sticky and began blistering. The once green needles were drying, falling, and forming a deep brown carpet on the ground. Overhead, a dark cloud, a thunderhead, blotted out the sun. Instantaneously, the air cooled by a few degrees and everything was still: even the birds were silent. Suddenly, there was a brilliant flash of light, then a roar of thunder. A lightning bolt raked the old Lodgepole Pine and its sap burst into flame. The flame was fanned by the wind and the tree blazed. The carpet caught fire. The flames were whipped by the wind. The fire spread to the old pine's neighbors. Dry trees exploded as the wall of heat reached them, sending burning debris in every direction, consuming the tons of fuel that had built up over decades. The fire rushed along valleys and up to the ridgetops, creating its own wind. Firestorm!

The fires of 1988 raged for days. They devoured 989,000 acres of America's first and most revered national park, Yellowstone. Millions of trees were destroyed. In a nightmarish scene right out of Bambi, the mammals fled before the flames. The fire was so intense and spread so rapidly that the young and very old, the sick and the slow could not run fast enough to save themselves. Insects, the most abundant animal life in the forest, had no chance to escape, and millions upon millions were incinerated. Yet, the more mobile animals, the birds and mammals, did survive. The number killed was amazingly low. Night after night, the TV news was filled with pictures of flames leaping hundreds of feet into the air. Images of charred forest were seared into the American consciousness. The message was that this was a terrible "National Tragedy," a loss of America's national treasure. The public, led by the TV news media, looked for someone to blame. "Who was the villain who stole our national park from us?" The implication was that someone or something evil was among us. As Pogo said: "We have seen the enemy and they are us."

This fire was a natural disaster, but was it really a tragedy? From our human perspective it may seem to be a catastrophe, but let's look at the fire from the perspective of the forest ecosystem. Is fire solely evil, or is fire actually part of the natural life cycle of the forest? Can it be seen as a positive force? Was the Yellowstone ecosystem more in harmony with its environment before or after the fire?

Disturbances help generate the mosaic makeup of the habitat. A fire burns out a patch of forest and opens it up to sunlight. Now, small plants, which had been suppressed by the shade of the trees, can thrive, and then a meadow can develop. Every organism is uniquely adapted to a particular type of habitat and a diverse array of habitats can support many more species than a uniform habitat. A variety of habitat patches, in turn, supports a diversity of species and communities. This biodiversity is the foundation of the natural ecosystem services upon which all life depends. Contrary to common thinking, disturbances are not bad, but rather they are valuable—indeed, they are essential for healthy ecosystems. Even the Yellowstone fires brought some important benefits. They created beautiful and diverse meadows of wildflowers and revitalized the forests. The nature of nature is change.

All disturbances are not alike. Some, like wholesale human alterations of natural systems (e.g., dams that "permanently" flood whole river valleys), can be devastating to their ecosystems. In this book, I will address the issues of scale and intensity of disturbances and their effects. Then, armed with these insights, I will examine changes in ecosystem management, and in our own lives, that are necessary if we are to live in harmony with nature's changing rhythms.

How Do Ecosystems Really Work?

The idea that a forest fire is a disaster for the forest is grounded in a long ecological tradition. For a century, we have assumed that constancy is the natural order of things. This basic idea is called an *equilibrium model.* This model presents an entire structure of thinking about nature, a paradigm. In this view, communities and ecosystems are supposed to remain constant. Is this a realistic view? Do communities really stay the same or are they constantly, naturally changing?

A community is the collection of all the organisms that live together in a place. It is composed of all of the animals, plants, fungi, bacteria, and viruses in a known area. We can thus refer to the "forest community" of Yellowstone National Park. As we descend through the Grand Canyon of the Yellowstone to the Yellowstone River, we will find that the community of the canyon walls is different from the community of the high plateau, and the community of the river is even more different. The plateau has Lodgepole Pines, caribou, deer, wolves, voles, and grasses. In the Yellowstone River, the terrestrial plateau community is replaced by something completely distinct. The river community is composed of aquatic algae and reeds, mayflies, dragonflies, and fish. Most people, and all ecologists, recognize that communities change in response to changes in climate and environment.

In contrast, our attitudes about how communities respond to variations over time have been very different. We generally expect things to be constant, to continue in the same way, making a tacit assumption that nature is unchanging, constant, forever. We may want it to be true, but that's not the way nature works. Change is the only constant.

My wife and I recently discovered a waterfall in western North Carolina called Catawba Falls. We loved the sight and sound of the water's 150-foot plunge, with its multiple chutes and torrents. The trees alongside the falls were tall and erect. Hemlocks and pines made a green cloak for the silver, dancing cascade. A large dead log spanned the stream at the base of the falls, at the perfect height to make a bench for us to sit on and soak up the beauty of the scene. We went back to the same spot a year later. We found that many trees had been blown down in a storm and lay strewn like pickup sticks crisscrossing the formerly unblemished cascade. I was disappointed: my waterfall had changed and our "bench" was gone. I felt a sense of loss. Yet, as I stood watching the falls, the altered beauty of the scene slowly transformed my disappointment into a sense of wonder and admiration. This new and revised edition of Catawba Falls was just as beautiful, just as awe inspiring as the previous edition. And the former edition was not the first edition, either. These falls have been transforming for millennia.

This changing waterfall is the life-sustaining environment for the host of aquatic insects and mosses that live in that part of the river. As you look at any river or stream, you can see tremendous variety in the flow of the water. In the midst (and the mist) of Catawba Falls are hundreds of different flow rates. In some places the water is rushing, churning, and frothing into whitewater (the white color is caused by the air bubbles trapped within the water), yet only a meter away the water is perfectly still. The rushing water was diverted by a boulder. Sheltered by the giant rock, a pool of quiet water was formed. These different places, or microhabitats, are filled with different groups of organisms, each adapted through natural selection to live only there. In the fast-flowing chutes are aquatic mosses with net-spinning caddisflies clinging to them. The water carries food to their nets for them to eat. In slightly faster water, blackfly larvae sit, clinging to scoured boulders, with their filtering fans projected up into the flow to catch their dinner. In the quiescent pools, burrowing mayflies feed on the animals deposited in the sediments. At the surface, water striders perch delicately balanced on the surface film, hunting for smaller terrestrial insects trapped on the water's surface. Each of these animals is well suited ("adapted") to its own microhabitat, its own special place in the river. Where the confluence of all their requisite environmental factors occurs, that's where you'll find them. That is what ecologists call their *physical niche.* I use "niche" as in ordinary common usage; it is an organism's place or role in nature.

However, microhabitats and environments change. If a tree falls in the forest, and lands in the river, it will change all of the flow patterns near it. What was once a fast reach can be stilled. As the water seeks a downstream path around the log, a new chute will form. The microhabitats have all changed, and the community must change, too. What was once an excellent blackfly habitat is no longer suitable for the flies. The animals have to move to survive. Some won't make it. Some were crushed when the tree fell on them. Some will get stranded and die. Some will get up and go. Change is endless and has consequences for all living things. It is as obvious as day follows night and the change of the seasons. What is rare is constancy.

Yet, people are drawn to an image of nature—a model—which assumes a constant environment. We can justify this obviously false assumption because we are overwhelmed by nature's complexity. An ecologist, working even in a rather simple community, must consider dozens of species, all interacting with one another. Some compete with each other for food, while some are food for each other. Just mapping out the feeding relationships, the food web (who eats whom), among the species in a community is a daunting task. Because in all communities, as in Catawba Falls, environmental change (e.g., from a simple treefall to a major disturbance) can rearrange all of the relationships, it is not surprising that ecologists have sought a way of simplifying the complexity.

Ecologists have often assumed that the environment is constant. The equilibrium model starts like old school math proofs: "All things being equal." We know now—as we suspected back in high school—that all things are never really quite equal. Ecologists have built growth, change, and evolution into the equilibrium model of the community. The process of community change is called *succession.* Succession is viewed as an orderly, sequential process, and its endpoint, the culmination, is called the *climax community.* Each stage, however transitory, can be treated as an intact community. The climax community is the ultimate, the Platonic ideal community for that environment, "the way things are supposed to be." This traditional, orderly view of nature requires that you have a constant environment to allow the community to reach the climax stage, that is, to equilibrate. This ideal climax community can become a reality only if the environment is actually constant. If the environment is in perpetual flux, then the climax community becomes impossible.

The traditional (equilibrium) perspective also dictates how one views the forces that structure natural communities. If you believe that the proper state of nature is to be in balance, then natural disturbances can be dismissed as aberrant. In contrast, from the newer, nonequilibrium point of view, disturbances are very important agents for promoting biological diversity.

To understand this dilemma, we need a few clear definitions. *Community structure* is the term that describes how the community of organisms of a given area (a stream reach or a forest tract) is organized. The key idea in community structure is biodiversity. The most basic question about the structure of any community is: "How many different species are there in the community?" To answer that question, we sample the community and count up the different species, a measure called the *species richness.* If you list all the names of all the species in the community, species richness is the tally at the bottom of the list.

In practice, this species list is abbreviated. In many communities we don't know everyone's name. There are thousands of species of microarthropods (e.g., mites and springtails) in the soil, too many to identify. The number of species of bacteria in soil or in water are myriad and extremely difficult to identify. So, in most analyses of the community, only the well-known and accessible groups are identified and tallied. For example in the Yellowstone mountain forests there are only eight common species of trees, about sixty-five species of birds, and about twenty-five species of common mammals. However, we have hardly any idea how many species of bacteria there are. Sometimes, people refer to local groups of similar species as if they were whole communities, for example, the tree community or the bird community of a forest, when the whole living forest (trees and shrubs, birds and bugs) is the community.

Biodiversity is more than species richness. It includes, among other things, the differences among individuals in populations, the differences among populations, and the relative abundance (or distribution) of individuals among species in a community (the evenness). For now, let's stick to species richness as our primary measure of biodiversity. Communities with higher species richness are more diverse than communities with fewer species. The high biodiversity of the macrobenthic invertebrates (macro = big; benthic = bottom dwelling; invertebrates = animals without backbones, mostly insects) of streams stands in striking contrast to the relative paucity of macrobenthic species in ponds. Ponds have fifteen to thirty species, while streams have nearly ten times as many. Why should neighboring systems with similar, even evolutionarily related groups of organisms (with overlapping families and genera), have such different community structures? What can explain these dramatic differences in biodiversity among similar communities? Understanding the underlying causes of differences in biodiversity is the central question of community ecology. As the great ecologist G. Evelyn Hutchinson asked in 1959, "Why are there so many kinds of animals?" The same question can be asked about plants as well. We are still striving to answer this most fundamental question. Let me give you a preview of where I am heading: to show that disturbance is emerging as a key force in creating and maintaining biodiversity.

What Determines Community Structure?

Historically, ecologists have been divided about what factors are most responsible for determining the structure of the community. The debate began early in the 1900s, when the basic issue was (and still is) whether the most important factors are biological or environmental. Are communities structured by the interactions among the organisms (e.g., competition, predation, etc.), or does the physical and chemical environment set the array of organisms? At its simplest level, the community is composed of the organisms that are able to live there. We don't find cacti in the rain forest, or tropical mahogany trees in the Rocky Mountains. One view is that the environment sets the community composition, meaning its membership and the relative abundance of its members. The presence or absence of a given organism in a given place is based on the ability of the organism to get there, to survive there, and to reproduce there. The physical and chemical environment sets the range of conditions for colonization, survival, growth, and reproduction of all possible residents of the community.

However, for any environment, the membership in the community is not fixed. There are far more potential species that can live there than are actually found there. Consider your choices when you plant your garden. When you go to the garden shop or the farmers' market, you can choose from a wide selection of annual plants. You can plant your garden with petunias (many varieties) or marigolds (all sizes and shapes and colors).

You make your selection for your garden from among dozens of species and hundreds of varieties. You are the selective agent, determining the species composition of your garden, the structure of the community of annual plants. Charles Darwin, in the *Origin of Species*, taught us that there were more individuals and varieties of a species than could survive and reproduce. Similarly, there are more species than can fit into any environment. He offered the idea of natural selection (or survival of the fittest) as a general explanation of the species that we see. Community structure follows the same pattern: it results from sorting out the various potential occupants of an area. How does this work in nature?

Charles S. Elton, one of the first great ecologists of the twentieth century, emphasized the role of interspecific interactions in determining the "limited membership" of the community. Interspecific interactions refer to the whole range of contacts, such as competition and predation, among different species. Elton did a clever thing. He examined the published surveys of fifty-one animal communities from all over the globe and from twenty-one different habitat types. He tallied the number of genera (groups of closely related species) and the number of species in each of the fifty-one communities. (This analysis makes the typical assumption that members of the same genus are more ecologically similar than species from different genera, which is usually a safe bet.) Then Elton computed the average number of species per genus in each community. He found that in his fifty-one surveys there were, on average, only 1.38

species per genus in each of these communities, or just a fraction more than one per genus. Then he compared this number to comprehensive taxonomic lists of species per genus in several well-studied groups across a wide variety of habitat types and communities (notably insects from all habitats in the United Kingdom). He found that, on average, across a range of communities, there were 4.23 species per genus.

Elton concluded that any given habitat has far fewer species per genus than are available in the larger world. Membership in the community was limited to about one species per genus, so only a small fraction of the total number of species that could potentially colonize an area actually coexists there. Why should this be? Elton's argument was that the interactions among the species, especially competition, determine community structure. He argued that these competitive interactions should be more intense among the species of the same genus, since they are morphologically and ecologically more similar than representatives of different genera. Elton's results can be generalized to conclude that biotic interactions determine community structure and biodiversity. Competing species contest each bit of turf. Predators ("red of tooth and claw") stalk their prey and eat them. Elton viewed the environment merely as the backdrop against which the real drama of the interactions among species is played out.

In this scenario, the environment is viewed as static and passive. If it varies at all, then the impact of the variation is considered trivial compared to the consequences of the interactions among species. The presumption of the preeminence of biotic interactions is not so strange, since ecologists are, after all, people. We are all drawn to the dramatic moment. Scenes of animals struggling together, fighting, or preying on each other is the main attraction of nature programming on television. Competition and predation attract ecologists, too. Mutualism, the study of how organisms help and benefit one another, gets very little press. Studies of mutualism account for less than one percent of the number of studies on either competition or predation.

The long-held assumption of environmental constancy is also understandable, since communities superficially appear to be constant and stable. The trees that were there yesterday are generally still there today. Yesterday's weather is a fine predictor of today's weather. Elton helped set the stage for ecologists to minimize the influence of the environment on community structure.

This equilibrium model of community structure is based on the interactions among species, and each species' relative strengths and abilities. Since the outcome is predictable, determined by the players, their characteristics, and their relative abundances, it is called a *deterministic model.* Much of traditional ecological theory is structured this way. Mathematical models predict the numbers of individuals of the competing species or the predators and their prey, given the species, their characteristics, and their relative abundances. Once you start building ecological models, which predict the future community structure, the models get very complex very quickly. In order to limit the complexity of these models, the first simplifying assumption that is made is to assume that the environment is not changing. We solve the equations for that moment when the abundances of all the individuals of each species are neither growing nor declining. That is, we solve them at equilibrium. This body of theory has structured ecologists' worldview for nearly a century. It can be called the *equilibrium paradigm.* It is the core concept organizing how nature is viewed by many people, often called the "balance of nature." The equilibrium paradigm has subtly guided how we manage our environment. It is the underpinning of the Smokey Bear syndrome of suppressing forest fires and is the conceptual framework behind the construction of dams for flood control. In short, it serves as a guiding principle of how the world is "supposed to be." It is loaded with value judgments, for example, that climax communities are good and disturbed communities are less desirable or somehow spoiled. However, as Bob Dylan sang, "The times they are a changing."

A new paradigm about how nature works has been emerging. Over the last two decades, an alternative worldview to the equilibrium model has gained strength in ecology. This is the concept of nonequilibrium dynamics of communities and is at the core of disturbance theory. In this view, communities are commonly disturbed, whether by fire, storm, drought, flood, or earthquake. Disturbance theory takes into account the changes that "disasters," both great and small, bring about in the natural world. Disturbances create new habitats and new opportunities for species to thrive, enhancing biodiversity. In this new paradigm, the normal state of the community can be thought of as recovering from the last disturbance, with the only constant being change. The goal of this book is to explore the meaning and consequences of this paradigm shift. If the environment is not constant, and if nature is not in balance, then where do we stand? How do we approach a world that is not now, nor ever will be, in equilibrium? The crux of this debate is whether interactions between species or forces of the environment (disturbances in particular) control community structure.

Nonequilibrium Determinants of Community Structure: The New Paradigm

The realization that real environments are constantly changing has renewed the fundamental debate in ecology over whether natural systems are dominated by the interactions among species or, alternatively, by environmental fluctuations and perturbations. In the 1950s, the big ecological debate was over how single-species populations were regulated, that is, why population sizes tend to stay within seemingly proscribed bounds. Now the focus is on what determines the structure of entire communities.

In the late 1960s and 1970s, the equilibrium model of community dynamics was crystallized by a group of ecologists led by Robert MacArthur. MacArthur and

E. O. Wilson's *The Theory of Island Biogeography* (coincidentally also published by Princeton University Press) is the epitome of an equilibrium model of community structure. They argued that the species richness on an island is controlled by the trade-off between immigration and extinction; the only role for the environment is to provide a set of resources. The only place the environment was included was as the size of the island. The environment was considered only a static "fruit bowl" of resources. Neither variation in space or time nor variation in the abundance or availability of resources was considered. The model argued for the absolute primacy of species interactions. In the theory, the solution to the number of species on an island is obtained when immigration and extinction balance exactly—when they are at equilibrium. It is noteworthy that this model of island biogeography still is widely used to predict the optimum size of nature reserves (such as national parks) to preserve biodiversity.

As is clear by now, equilibrium models presume a constant environment. Such models exclude disturbances, and any other environmental fluctuations. In this view, it is a simple step to conclude that biotic interactions are the key determinants of community structure. Under equilibrium conditions, the community is the direct result of the competitive and predator-prey relationships among and between species. The environment is viewed as predictable, regular, and constant.

In all of these equilibrium ideas, the environment is the backdrop, not the actor. It is as if the waterfall at

Catawba Falls were just so much scenery and the real actors were the animals and plants and bacteria and fungi, living on and among the rocks. What's missing here is the understanding that the species only exist there because the waterfall suits them and the flow meets their needs, creating the conditions for successful survival and reproduction. If a drought dried up the Catawba River, nearly all the animals would die, and species richness would nosedive. In the period between disturbances, significant competition or predation may well be occurring, so that a quasi-equilibrium begins to become established. Yet when the next storm-driven flood hits, all bets are off. Surviving the flood becomes everyone's first and only priority.

Reconsider the Yellowstone fires. Walt Disney had it right in *Bambi.* The environment was ablaze. It was the fire that made the animals run. In the film, we saw predators and prey all mixed together, fleeing before the flames. Foxes and rabbits ran away from the heat, together. When the great fire sweeps through the forest, no one stops to eat; all notions of competition and predation are gone. Survival is the order of the day. Disturbances change all the rules. After the fires cooled, the entire community of Yellowstone was restructured. The environment was anything but passive. It played a pivotal role, as it always does. This applies not only in Yellowstone, but everywhere.

As the Catawba Falls and Yellowstone examples illustrate, the concept of a natural ecosystem at equilibrium is only an ideal, a vision of a world where species interact with one another, unconstrained by pressures from their environment. It is also a way by which ecologists, as biologists, have elevated the importance of biological interactions as determinants of community structure. Note that disturbance theory does not ignore competition and predation as important factors in community structure; it just puts them in perspective with the other vital forces of nature.

Are disturbances common and important enough to finally dethrone the equilibrium paradigm? They certainly are. Disturbances are common in all ecosystems. Table 1.1 gives a listing of disturbance frequency and predictability in natural ecosystems pointing out the major disturbance types that impact the world's ecosystems. It is not exhaustive, and you may be able to think of others. Some of these disturbances will be discussed in detail in chapter 2.

A false friend has fostered our attitudes toward disturbances. Smokey Bear was a National Forest Service creation born of equilibrium thinking about community dynamics and ecosystem management. Smokey taught generations of people that forest fires were bad, and that "Only *you* can prevent forest fires!" However, Smokey got us into trouble. In Yellowstone, management policy was to suppress all forest fires. Without fire suppression, there would have been many smaller fires over time, reducing the fuel load (the dead trees and branches) and lowering the density of trees. Since

Table 1.1

Disturbance Frequency and Predictability in Natural Ecosystems

Ecosystem Type	*Disturbance*	*Frequency*[*]	*Predictability*
Terrestrial			
Deciduous forest	Fire	1/40–200 years	None
	Windstorm	1/10–25 years	None
	Insect defoliation	Rare	None
Coniferous forest	Fire	1/20–40 years	Moderate
	Windstorm	1/10–25 years	None
	Insect defoliation	Rare	None
Rain forests			
Tropical	Windthrow	Frequent	None
	Fire	Frequent	None
Temperate	Fire	1/200–500 years	None
	Storms	1/50–100 years	None
Chaparral	Fire	1/15–25 years	High
Grasslands	Fire	1/5–10 years	Moderate
Desert	Frost	1/50–200 years	None
Pocosins	Fire	1/10–25 years	None
Freshwater			
Streams and rivers	Floods		
	Spring snowmelt	Annual	High
	Storms	0–15/year	None
	Drying up	0–2/year	Moderate to high
	Freezing/Anchor Ice	0–2/year	High
Lakes	Storms	0–4/year	None
	Freezing (Winter Kill)	0–1/year	High
Ponds	Freezing	0–1/year	High
	Drying up	0–2/year	Low to high

TABLE 1.1 (cont.)

Ecosystem Type	*Disturbance*	*Frequency**	*Predictability*
MARINE			
Intertidal zone	Hurricanes	1/20 years	Low
Beaches	Tornadoes	1/20 years	Low
	Log damage	Annual	Low
Pelagic zone	Storms, hurricanes	Aperiodic	Low
Deep-sea benthos	Storms, hurricanes	Aperiodic	Low
	Submarine volcanoes	Aperiodic	Low
	Whale carcasses	Aperiodic	Low

* Number of disturbances/unit time

smaller fires had been suppressed for decades, the fuel load in Yellowstone in 1988 was immense, and as a result the 1988 fires were devastating. The timing of the fires was unpredictable, although the conditions that allowed the fire to start had been building for generations. The destruction wrought by the 1988 fires would have been far more limited if previous natural fire disturbances had run their course.

What was the real threat to Yellowstone, then? Strange as it may seem, it was not the fires. The fires opened up the forest and allowed many trees to germinate for the first time in decades. The fires cleared out the deep litter, exposing soils to sunlight and allowing many wild-

Figure 1.1 The Yellowstone fires of 1988 created a patchy mosaic of regrown and unburned areas. (Photo by S. R. Reice, 1998)

flowers to bloom and reproduce. The fires re-created the patchy mosaic landscape of burned and unburned stands, of meadows among the forests (see fig. 1.1). The real threat to Yellowstone was the fire suppression policies of the National Park Service, which created the conditions for this conflagration in the first place. This may seem counterintuitive, but this book will argue that the absence of disturbance, not the disturbance itself, is the real danger. Communities and ecosystems require disturbances for their very survival. In this book we will take a closer look at the positive role of disturbances in nature and show that they are vital to maintaining the integrity and health of natural ecosystems, upon which all life depends. This is the silver lining in the storm cloud we must learn to seek out and value.

Further Reading

Botkin, D. B. 1995. *Our Natural History: The Lessons of Lewis and Clark.* New York: Grosset/Putnam.

Darwin, C. 1988 [1859]. *On the Origin of Species.* New York: New York University Press.

Elton, C. S. 2000 [1927]. *The Ecology of Invasions by Animals and Plants.* Chicago: University of Chicago Press.

MacArthur, R. and E. O. Wilson. 1967. *The Theory of Island Biogeography.* Princeton, NJ: Princeton University Press.

Quammen, D. 1997. *The Song of the Dodo: Island Biogeography in an Age of Extinctions.* New York: Touchstone Books.

Chapter 2 ***Disturbance, Patchiness, and Communities***

When written in Chinese, the word "crisis" is composed of two characters. One represents danger and the other represents opportunity.

—John F. Kennedy, April 12, 1959, Indianapolis, Indiana

Chapel Hill, North Carolina, September 5–6, 1996

During the night we awoke to howling winds and the piercing "crack!" of trees snapping outside our bedroom window. Then, it was quiet, and sleep came at last. In the hot, humid dawn, we awoke and saw the unthinkable damage. Trees were down. Debris was everywhere. By portable radio we learned the name of this nocturnal scourge: Hurricane Fran. As we began to recover from the initial shock, we learned the extent of the damage wrought by the storm: one million people without power; roads closed; buildings demolished; twenty-four human lives lost; $2.3 billion worth of property destroyed. Our hold over nature, which had seemed so powerful, had been broken overnight.

I went out to survey the damage. Our house had been spared. However, behind our house, and down into the floodplain of Neville's Creek, over one hundred trees—more than 50 percent—were down. Massive Loblolly Pines were uprooted, and the trunks of smaller ones had snapped. Two-hundred-year-old oaks and hickories were flattened. Neville's Creek, a normally quiescent trickle of water was choked with downed trees and roaring with whitewater from all the runoff. The few trees that still stood were leaning at bizarre, threatening angles. The devastation was incredible (fig. 2.1).

I went around to the front of the house. No damage—none! Not even one tree was down. Impossible! How could this place, only 150 feet (50 meters) uphill from the destruction along Neville's Creek, have escaped the storm? What could account for these grotesquely different impacts of the same hurricane? As the roads were cleared and became passable again, I could see the extreme patchiness of Hurricane Fran's impact. One neighborhood was virtually untouched and another one smashed. Trees crushed one house but didn't touch its neighbors. Was Fran uniquely capricious, or is this variation in damage typical of natural disasters and disturbances?

On closer examination, a pattern started to emerge. The storm had caused far more damage along streams and rivers, in the floodplains and down in the valleys,

Figure 2.1 The aftermath of Hurricane Fran, Orange County, North Carolina, 1995. (Photo by Rickie White)

than on higher ground. Why? In the week prior to Hurricane Fran, eastern and central North Carolina had gotten heavy rains, up to 5 inches in some places. Thus, the soils in the floodplains were already saturated when Fran dumped up to 9 more inches. In these soggy soils, the tree roots were already loose, making the trees in the valleys more susceptible to the storm's winds. The older, tall hardwoods, whose dense, leafy crowns acted like sails, were easily blown over. They were ready to fall. When the winds of Hurricane Fran came, these were the casualties.

The hurricane itself was not uniform in its destructive potential. Its very nature was patchy. The winds in the eye of a storm are nearly calm. The winds 150 miles inland had peak gusts of more than 75 miles per hour in some places, while others never exceeded 25 mph. The patterns of destruction actually make good ecological sense. Amid the immediate destruction of Hurricane Fran, it was hard to imagine any benefits from it. Was there any silver lining here?

What Is a Disturbance?

No community has a truly constant environment. Whether it is the change from day to night or the march of the seasons, we scarcely give these changes any thought. They are normal. Disturbances are changes, too, but ones that are out of the ordinary. These can be the result of severe weather, like Hurricane Fran, or forest fires, floods, or volcanic eruptions. It is rather hard to define what, precisely, is a disturbance, but here is a straightforward definition: "A disturbance is a physical event with biological consequences." A more complete definition is that given by ecologists Peter White and Steward Pickett, who wrote: "A disturbance is any relatively discrete event in time that disrupts ecosystem, community or population structure, and changes resources, availability of substratum, or the physical environment." I prefer this definition because it defines disturbance as the physical event, which alters the biota (the living organisms). It is tempting, but misleading, to define a disturbance as the biological outcome of the event. Here's the difference. Say a hurricane knocks down 10,000 trees. The dead trees are not the disturbance, but the result of the disturbance, the hurricane. I view a disturbance as a physical force, such as a fire, flood, tornado, or hurricane, which damages natural systems and removes organisms. Here is a simple rule of thumb: the initial impact of a disturbance will always be the removal of organisms; if no organisms were removed or killed, the event was not a disturbance.

Our understanding of disturbance is dependent on the scale: *temporal scale* is the length of time, and *spatial scale* the area being observed. If the area of observation is large enough or the period of observation long enough, all disturbances are predictable and "normal." That is, if we look at all forests around the globe for a period of a year, at some place and at some time there will be a fire burning. However, when we shrink the scale and look only at a particular small woodlot for just a day, the disturbances appear completely random. We will never know precisely where or when a particular fire will occur. So, when considered at large scales, disturbances are common and predictable. Yet for any particular place and time, disturbances appear to be random. Since every place eventually gets disturbed (if we wait long enough), we can state as a general rule that all communities are recovering from the last disturbance.

If you keep this principle in mind when you are walking outdoors, you will soon learn to recognize the signs of the previous disturbances. You might see fire scars or charred bark on healthy trees, clues that a forest fire burned here. You often see mounds of dead leaves resting several feet off the ground in the limbs of trees beside a river. This is a sign that a flood raised the leaves that far above the ground, trapping them up in the branches.

If a destructive event is fully predictable, the organisms can and will adapt to it. A disturbance that is unpredictable will, therefore, have a greater impact than one that is relatively predictable. For example, predictable spring floods wash the fine silts out from between the pieces of gravel in a gravel bed in mountain streams. This makes the gravel bed more suitable for trout and salmon to build their nests and lay their eggs. Oddly enough, the removal of predictable "disturbances" (e.g., preventing annual spring snowmelt-driven floods or preventing forest fires) often has a greater, more negative impact on the ecosystem than a "normal," predictable flood or fire. If the spring flood doesn't occur, the buildup of silt in the gravel bed will diminish the spawning of fish. On a personal and human level, I grew up on a noisy street in New York City. When I first went to college, it was so quiet that I could not fall asleep. I had acclimatized to the predictable noise (disturbance). Eventually I did get used to the quiet.

Contrast this view of disturbance to the effect of predation. Predation is an important agent for the removal of prey individuals. Is it really the same as a disturbance? Predation is intrinsic to the life of the prey organisms, and the prey species adapts to this. Disturbance, however, is "a punctuated killing." Species do adapt to predictable environmental regimes (such as late-spring, snow-melt driven floods in the Rocky Mountains). Although there is an ongoing debate on this point, I believe that if a flood is predictable, it should not be considered a disturbance.

Toxic spills, which kill many individuals and species, clearly fit the definition of disturbance. When the *Exxon Valdez* accidentally dumped its oil into Prince William Sound in Alaska in 1989, beaches, birds, and sea otters became coated with the gooey substance. Many species suffered major population losses. Organic waste spills (e.g., from ruptured hog-waste lagoons) are not toxic in themselves, yet large amounts of organic matter added to a stream can cause a fish kill. Bacteria break down and decompose organic matter while using oxygen to respire. When the hog manure hits the stream, the bacteria grow and reproduce. Each new cell requires yet more oxygen. The added organic matter creates a high biological oxygen demand (BOD) in the water. Shortly, the bacteria use up all the oxygen in the water, and the fish suffocate. The manure spill was the indirect cause of the death of hundreds of thousands of fish. The fish kill was actually caused by the loss of dissolved oxygen (DO). What was the disturbance? Even though the spill was not toxic in itself, it still counts as a disturbance because it resulted in a fish kill.

Disturbance effects are typically not limited to one population at a time. When the habitat is altered or rearranged by a disturbance, organisms are removed. Disturbances can cause restructuring of a whole ecosystem. Natural communities, whether fields or forests, lakes or streams, go through a series of ordered stages of development that ecologists term *succession.* For example, an abandoned field in the southeastern USA is first colonized by annual plants; then perennial plants; then shrubs; then pines; and then oaks. A disturbance, for example a brush fire, can reset the entire community back to an earlier stage of succession. We call this *reinitialization,* since the successional development of a community must start over, often from the beginning.

Different disturbances will remove different species to different degrees. Therefore, the response of the community to any particular disturbance is likely to be a unique successional sequence. If a forest community had been undisturbed for hundreds of years and was approaching the steady-state (climax) condition, a disturbance would reset the community back to an earlier stage. It can also result in a community with an entirely new species composition that had never before existed. The forest community in Piedmont, North Carolina, that now exists after Hurricane Fran is brand new. We suddenly have a forest community with far fewer mature, large oaks and hickories in it. How will the succession of this new community proceed from this point forward? What will the new tree species' composition be next year, or fifty years from now? Only time will tell.

Disturbances Are Not Only Common, They Are Often Essential

Disturbances are common and important in virtually all ecosystems (see table 1.1). For example, in terrestrial ecosystems, fire and windstorms are the most important disturbances. Neither is predictable, although fire is more typical in the summer and autumn months, but both can destroy plants and animals. Let's examine a few terrestrial systems that are adapted to fire or are dependent on it.

Many species actually require fire to flower or to produce new seeds. In chaparral (e.g., in Southern California and around the Mediterranean region), the life histories of low-shrub vegetation—species such as Manzanita and Scrub Oak—are linked to the periodic disturbance of fire. As debris builds up during non-burn periods, the susceptibility to fire increases. The reproduction of these plants is keyed to fire, as their seeds will not germinate without it.

Similarly, serotinous pine trees have cones that will not open until heated by fire. These include the Lodgepole Pine (*Pinus contorta*) of western North America, Jack Pine (*Pinus banksiana*) of north central North America, Longleaf Pine (*Pinus palustris*) and Pond Pine (*Pinus serotina*) of the southeastern USA. The buildup of debris on the forest floor enhances the probability of fire occurring. Not only is the release of seeds from cones fire dependent, but fire also uncovers bare soil and releases nutrients, both prerequisites for successful germination and growth. Without fire, these pines' reproduction would be halted by the accumulation of their own litter, and they would soon be replaced by other tree species.

In the Australian bush, *Eucalyptus* forests are fire adapted and fire maintained. This process is not limited to some particular species but applies to the dominant genus of an entire continent. These communities are populated by trees that are fire dominants, that is, they are dominant only because of fire. Fire dominance is the rule in Australia. In the Northern Hemisphere, fire dominance is relatively rare today, though it was common in presettlement vegetation.

In stream communities, disturbance by flooding plays a major role in maintaining high biodiversity among the animals. Stream communities change dramatically after floods. A disturbance eliminates some species and favors others. Benthic (bottom-dwelling) invertebrates, in flowing waters that flood frequently, have adaptations that allow them to cling to bottom substrates such as stones, gravel, or submerged logs. Many benthic insects such as mayfly nymphs (Ephemeroptera) have claws that hook into minute crevices and cracks in stones. Others live in the protective, calm boundary layer, the still layer of water that results from friction between the flow and the substrate. Blackfly larvae (Simuliidae) cling to rocks. Their bodies remain in the boundary layer and are protected from the full force of the current. Blackflies have fanlike appendages that project upward into the flow to filter food particles. Blackflies are excellent colonizers of newly opened spaces on rocks created by scouring floods. Similar examples of adaptations of species to disturbance regimes are common in many systems.

Some ecosystems have disturbances that are less dramatic than fires or floods. Drought can be a significant disturbance in freshwater or terrestrial systems. Storm-induced mixing or pounding of shorelines by wind-driven waves can be a significant disturbance for lakes. Freezing and drying can severely impact pond communities. The deep ocean is less vulnerable than surface waters to storms. Yet, try to imagine a whale carcass hitting the ocean floor. The impact can cause localized enrichment while crushing many smaller individuals and rearranging the entire sediment structure. According to Anthony Jones of the Australian Museum, whale carcasses occur at a density of one per 3.9 square miles (10 sq km) on the Pacific Ocean floor! There are also periodic eruptions of submarine volcanos and submarine earthquakes. So, even the apparently calm deep-ocean bottom gets disturbed. All ecosystem types are disturbed, and most are disturbed rather frequently (table 1.1). The frequency, timing, and intensity of these disturbances is surely a driving factor in the evolution of the species in every ecosystem.

For an equilibrium community to be established requires an ecologically long disturbance-free period. Ecologists now recognize that this virtually never happens. Therefore, the normal condition for communities is that they are constantly recovering from the last disturbance.

A Brief Introduction to Disturbance Theory: How Do Disturbances Affect Biodiversity?

The direct effect of disturbance is to remove individuals from a community. They may be killed or transported away by the direct action of the disturbance. The dynamics of the ecosystem following the disturbance is what determines the community structure. This new era of thinking was spurred by a paper by the eminent ecologist Joseph Connell in 1978. He presented the Intermediate Disturbance Hypothesis (IDH) to explain the high species diversity in tropical rain forests and coral reefs. (Others had already suggested a similar view, but this paper crystallized the argument and made it accessible and popular.)

Here's how the IDH works. Assume a competitive hierarchy of species (i.e. Species A always outcompetes Species B, B outcompetes C, C outcompetes D, and so on). At low frequency and/or magnitude of disturbance, the superior competitors will predominate. Without disturbance, they can increase in size and number to the point that they can eliminate the inferior competitors. This is called *competitive exclusion.* The removal of these inferior competitors reduces the species richness (biodiversity) of the system (see fig. 2.2, the

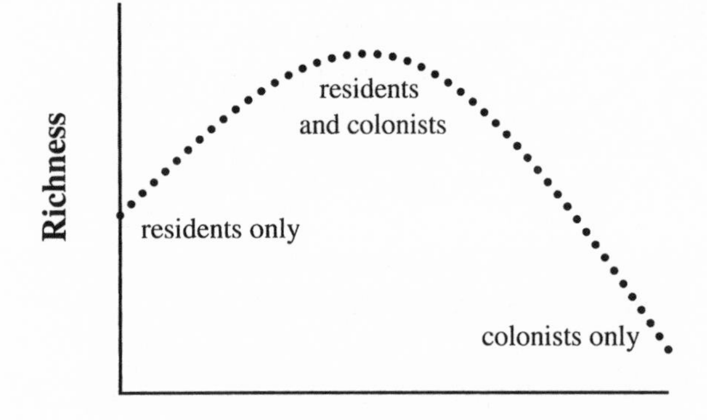

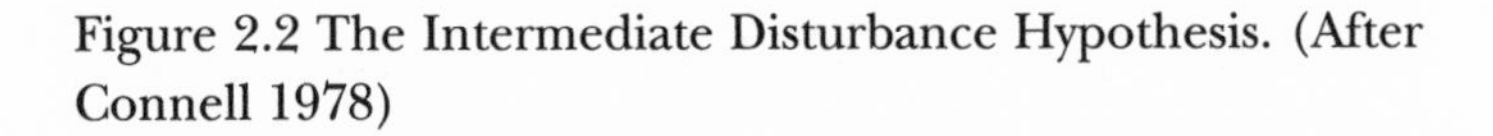

Figure 2.2 The Intermediate Disturbance Hypothesis. (After Connell 1978)

low species richness values on the lower left side). The model also presumes that the superior competitors are efficient occupiers of space. Let's call them resident species. These resident species are the dominant trees in the rain forest and the dominant corals on the coral reef.

What happens when disturbances are either frequent or severe? Then, the dominant competitors—along with nearly every other species—are reduced or even eliminated. In their absence, space is opened up for colonizing species that have high dispersal abilities. They arrive quickly after a disturbance, but are typically poor competitors. Some examples of these colonizers include crabgrass, which seems to recolonize your newly tilled garden almost as soon as you look away; *Cecropia* trees, which colonize gaps caused by treefalls in the rain forest; and blackfly larvae that colonize scoured rocks in streams. Under extreme disturbance regimes, these colonizing species will dominate the system, since they are the only ones that have time and dispersal abilities sufficient to become established. Then, the absence of the resident species reduces species richness (see fig. 2.2, the low species richness values on the lower right side). When disturbance is intermediate in frequency, magnitude, and/or intensity, some resident species will persist in the system. They will coexist along with colonizing species, which exploit the disturbed areas. Therefore, species richness peaks at moderate levels of disturbance (fig. 2.2, middle). This is the essence of the Intermediate Disturbance Hypothesis.

Ecologist Mike Huston's "dynamic equilibrium" model further developed the logic behind the IDH by focusing on the mechanisms that drive it. Huston showed that community structure can result from the trade-offs between population growth rates, rates of competitive exclusion, and frequency of population reductions. Huston demonstrated that if the return interval of disturbance was short (relative to the time necessary for competitive exclusion to occur), then poorer competitors would persist, thus increasing species richness. He wrote, "Diversity is determined not so much by the relative competitive abilities of the competing species as by the influence of the environment on the net outcome of their interactions." Disturbance is the main source of these population reductions. Since nearly all ecosystems are disturbed frequently enough to prevent competitive exclusion, high biodiversity is the unexpected result of disturbances and so-called disasters. Tornadoes and hurricanes do indeed have dark storm clouds, but this increase in biodiversity is their silver lining.

Patchiness and Spatial Heterogeneity and Disturbance

The ecological term for patchiness is *spatial heterogeneity*. All environments are heterogeneous. Terrestrial ecosystems vary in elevation, temperature, slope, moisture, and soil chemistry. Aquatic ecosystems vary in depth, temperature, turbidity, current velocity, salinity, and nutrients. In the complex habitats like rain forests and coral reefs, heterogeneity is immediately obvious to anyone (fig. 2.3). Even the most superficially homogeneous systems are heterogeneous. An agricultural field has raised rows (somewhat warmer and drier) and incised furrows (cooler and wetter). Fields have gradients in soil texture, moisture, and chemistry. Salt marshes, with only a single species of grass, *Spartina alterniflora*, has patches of sand and mud, gradients of salinity and moisture, tidal creeks, etc. The list of physical/chemical variables that change spatially is endless. Many ecological modelers continue to assume the existence of a homogeneous environment. This makes ecological models of populations, communities, and ecosystems far simpler to deal with, both mathematically and conceptually. Our best-designed field experiments can test the effects of only a few environmental variables at a time and do not begin to approach the richness of the environmental complexity that actually exists.

Natural disturbances are patchy largely because the landscape they act on is itself patchy. There is a complex set of interactions between the heterogeneity of landscapes and the distribution of disturbance effects. Forest fires seem random, burning some areas to the ground, yet barely touching adjacent patches (see fig. 1.1). Likewise, the chaotic impacts of Hurricane Fran, apparently random, are explicable. All natural phenomena have real causes. When things look as if they are random, it's usually because we simply don't fully understand what is going on.

In 1945, Charles Elton made the following observation: "There really is no such thing as a uniform habitat, since all habitats consist of interspersed mosaics of micro-habitats or are internally patchy in the distribution of population densities . . . and since they are also subject to variations in conditions caused by seasonal and other temporal changes." This wise man saw what we often ignore even today. The environment is endlessly heterogeneous in both space and time. Further, he implied that patchiness is due, in part, to the action of disturbances. Spatial heterogeneity paired with disturbance can generate and maintain high biodiversity.

(a)

(b)

(c)

Figure 2.3 Spatial heterogeneity on different scales. (a) Tropical rain forest. (b) Sweet-potato field. (c) Salt marsh (with the salt-marsh cord grass, *Spartina alterniflora*). (Photo by S. R. Reice)

Different Patches Have Different Vulnerabilities to the Same Disturbance

Disturbances seem capricious. A fire that burns one patch of woods but not another causes a patchy distribution of trees. A tornado touches down in a neighborhood, leveling one house, but leaving the house next door untouched. Fires and floods and tornadoes are simply not uniform across the landscape. Where they hit, they create patches. A big part of the patchiness that we observe in nature is due to the direct effects of disturbances. Yet, the underlying patchiness of the landscape influences the intensity of the effect of a given disturbance.

In all cases, some parts of the landscape are more vulnerable to one particular type of disturbance than another. Different habitats can moderate or intensify the magnitude of the disturbance event. Imagine a unit disturbance, that is, a disturbance of uniform strength or magnitude across the range of environments. The unit disturbance could be a wind of 100 miles per hour, or a fire of 300°F. Different patches will have differential susceptibility to that unit disturbance. The disturbance will enhance the underlying patchiness, since one patch will sustain far greater population losses than another. As a result, different patches will be open to different amounts of recolonization. Let's look more closely at this issue of varying vulnerability in some specific cases.

Example 1. Fire Susceptibility in Forests

Fire susceptibility in forests is a function of many variables. The main factors are fuel quantity, fuel quality, and moisture. If there is more fuel (dead wood), plenty of tinder (fine branches or needles or bark), and the fuel is dry, the probability of a forest fire increases. Forest rangers often issue warnings of high fire risk and may institute bans on outdoor burning. However, there are many other factors that determine the probability of fire at a given forest site. Soil moisture reduces both fuel quality and quantity. Wet wood does not burn well, and decomposition on the forest floor is more rapid where the soil is wetter. Wind is another important factor. Wind blows down trees and helps dry the fuel wood (slowing the decomposition rate), thus increasing both

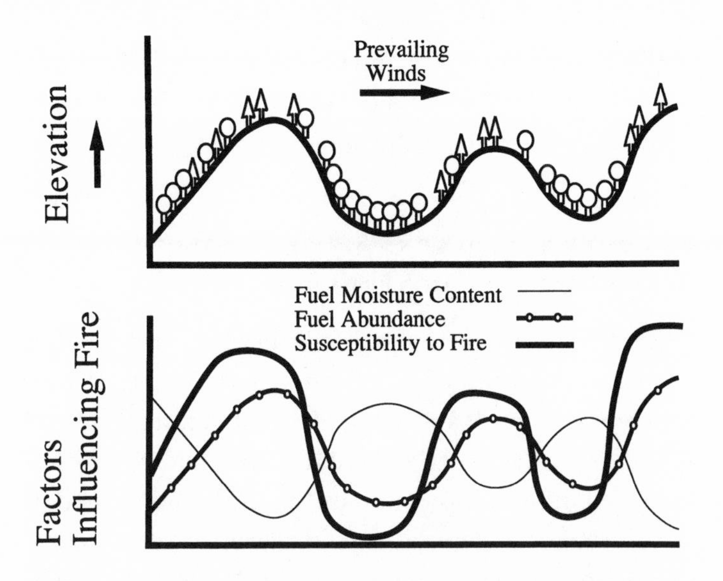

Figure 2.4 Fire susceptibility in the mountains is a function of fuel quality, quantity, and wind. (Illustration by S. R. Reice)

the fuel quality and the quantity. Wind also helps spread the flames once a fire starts. These factors combined are illustrated in figure 2.4. The susceptibility to fire is greater in the windy upper slopes than in the sheltered lower slopes. It is greatest on the mountaintops, and lowest on the moist valley floors. So, the risk of fire is not uniform across the forest. The intensity of the fire, once it starts, will also be variable in space and time. Variability in distribution and intensity of burns is common in all terrestrial systems. In the Yellowstone fire of 1988, whole forest blocks were burned to the ground while others were barely singed (fig. 1.1). In the Great Smoky Mountains National Park, the frequency of fire depends on elevation and soil moisture. The variation in fire frequency reflects the relative susceptibility of the different patches to the different sources of fire such as lightning strikes and anthropogenic fires.

Example 2. Sediment Size and Frequency of Disturbance in Streams

Sediments in streams are patchily distributed, since they are sorted out by different current velocities. (Faster flow washes out silt and sand, leaving behind gravel and boulders.) Patches of different size sediments will be disturbed at different frequencies (fig. 2.5), determined by their relative susceptibility to different current velocities. Disturbance frequency will be highest for sands, since sand can move even in rather slow flows. Larger sediment particles start to move at higher and higher velocities—gravel, then pebbles, then cobbles, and finally boulders, which move only during very swift flows. As particle size decreases from sand to silt and clay (from 0.05 to 0.001 mm), faster and faster flows are needed to move the particles due to the enhanced adhesive properties of the smaller particles. Clay particles stick together. (This simple fact was discovered in the Bronze Age, and led to the making of pottery and the beginnings of civilization.) Only major floods, which are rare, can disturb the clays or boulders. So, in any flood, some sediment patches will wash out while others will stay put, with their resident community intact. For all streams, sediment distribution is a key factor in determining the frequency, severity, and pattern of distur-

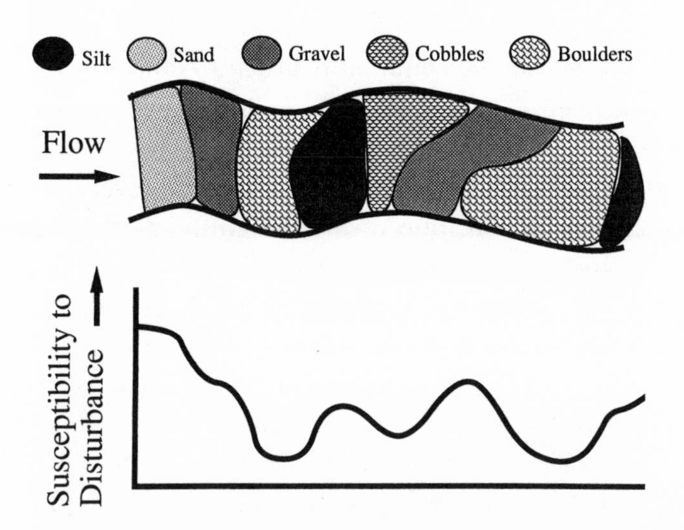

Figure 2.5 Different sediment patches in a river will have different susceptibilities to the current velocity and hence different frequencies of disturbance. (Illustration by S. R. Reice)

bance. Sediment movements can completely disrupt the stream benthic community. Invertebrates, incubating fish eggs, and attached algae can be scoured off, buried, or crushed by tumbling sediments, resulting in displacement, injury, or even death. On the other hand, this means that space is opened up, creating opportunities for recolonization.

The relationship among sediment particle size, frequency of disturbance, and community structure has not yet been thoroughly investigated. Such varied disturbance regimes should have direct consequences for the species composition and structure of biotic communities in rivers and streams. Resilience (the ability of a community to return to its original structure) of a sediment-specific stream community structure should be directly related to the frequency of disturbance in that sediment type. Higher frequency of disturbance should produce communities with higher rates of recovery. To test this prediction, one of my students, Dr. Jennifer Smith, and I imposed unit tumbling disturbances on patches of cobbles and patches of sand mixed with gravel. Species richness was reduced 24 percent in cobbles and 40 percent in sand/gravel. Total animal densities were reduced 48 percent in cobbles and 79 percent in sand/gravel. Within two weeks, these animal densities rebounded 240 percent in cobbles and 510 percent in sand/gravel. These data show that, in response to a unit disturbance, there are dramatic differences in space opened up and in rates of recolonization between two distinct stream sediment patches. This shows that a given disturbance will have different impacts on different patches in the same ecosystem.

Heterogeneity, Disturbance, and Recolonization

Spatial heterogeneity is frequently invoked as a mechanism for maintaining biological diversity. A more complex, variable environment allows species to subdivide resources, which we call *resource partitioning*. Beech trees thrive in wet bottomlands, while pines do better in drier upland sites. Coexistence of similar species is critical to maintaining high biotic diversity. How similar can the species be and still coexist? This is the traditional problem of *limiting similarity*, which goes back to Darwin. How can ten different species of oak (*Quercus* spp) coexist in the same forest, or why do twelve different species of mayflies coexist in the same riffle of a stream? Our problem here is that very similar species should be competing intensely for resources, and the best competitor should drive the others to extinction. The problem of explaining high species richness gets simpler when we recall that the environment is patchy, so that species can divide up the landscape, with each one living in the microhabitat where it has highest fitness. If similar species are favored in different patches, then competition is minimized or does not occur at all. This solves the problem of explaining their continued coexistence.

Yet, in nature we find many species combinations where the ecology of the species is not sufficiently different to explain why they coexist. Why doesn't competition simply eliminate the weaker species? Why isn't the best-adapted species (for a particular site) the only species that exists there? This is part of the Big Question of community ecology: *Why are there so many kinds of species?* My preferred answer to this question is based on the interactions among disturbance, patchiness, and recolonization. Disturbances open up different patches to different degrees, preventing competitive exclusion or predators eliminating their prey. Once these patches are open, then the process of recolonization can bring in new species, enhancing biodiversity (see chapter 3).

Whether the disturbance is a fire or flood, each distinct patch type in each system will have its own unique response. The heterogeneity of the landscape translates into different degrees of disturbance impacts on the community, resulting in and reinforcing the patchiness of distribution of species. The result of the disturbance is to create a mosaic of open spaces in the system and a range of opportunities for recolonization. Unimpeded, the forces of competition or predation can drive species to local extinction, reducing diversity in communities. However, disturbances can prevent competitive exclusion of poor competitors and prevent predators from destroying their prey species. When the forest is on fire or the river is flooding, all bets are off. If a fox is ready to pounce on a rabbit and it feels the heat of an approaching fire, then dinner becomes a secondary consideration, and the rabbit is free to flee. Disturbances interrupt these biological processes and suspend their power to eliminate species. So, disturbances take center stage as a mechanism to explain the existence of high biodiversity.

Let's go back to Hurricane Fran. Although the winds were strong, they were not equally strong everywhere and the devastation was not uniform. Because the environment is heterogeneous the winds were funneled through valleys while they were somewhat slower in more open, flat areas. Furthermore, some patches were more vulnerable to the disturbance than others. The impact and tree damage were greatest down in the bottomlands, where the soils were saturated. The newly opened patches, where the trees were knocked down, became available to colonizing species of trees and

Figure 2.6 Fireweed (*Erechtites hieracifolia*) in a light gap one year after Hurricane Fran, Orange County, North Carolina, 1996. (Photo by Rickie White)

herbs. The mosaic nature of the environment is preserved, which, in turn, fosters high biodiversity.

Now, years after Hurricane Fran, the Piedmont of North Carolina is still recovering. What changes do we see today? The mosquitoes are doing very well: the uprooted trees left holes in the ground, which filled with rainwater and became an ideal breeding habitat for them. Fran blew down trees, creating forest gaps, which favors poison ivy. In the new gaps there are enormous stands of Fireweed (*Erechtites hieracifolia*), 15 feet tall. Its crown of white feathery seeds makes a flying carpet, 10 feet above the ground (fig. 2.6). Red Maple (*Acer rubrum*) seedlings are everywhere, taking advantage of the reduction in competition from the mature oaks and the increase in light. Forest stands, which were heavily shaded and dominated by oaks and hickories, now support dozens of smaller, weedy species. Trees that fell into streams have re-created many pools, which favor native fishes and dozens of insect species.

Hurricane Fran reacted to the heterogeneity of the Piedmont and simultaneously created more heterogeneity, reshaping the landscape. The destruction of Hurricane Fran created opportunities for many species that had been suppressed or excluded. The big result has been an increase in the biodiversity of the area. In the fall of 1999, Hurricane Floyd wreaked more havoc and created forest gaps. In the wake of disaster, of danger, there is opportunity! Every cloud may indeed have a silver lining.

Further Reading

Connell, J. H. 1978. Diversity in tropical rain forests and coral reefs. *Science* 199(24): 1302–1309.

Gould, S. J. 1989. *Wonderful Life: The Burgess Shale and the Nature of History.* New York: Norton.

Huston, M. 1979. A general hypothesis of species diversity. *American Naturalist* 113: 81–101.

Levin, S. A. 1999. *Fragile Dominion: Complexity and the Commons.* Reading, MA: Helix Books.

Pickett, S.T.A., and P. S. White, eds. 1985. *The Ecology of Natural Disturbance and Patch Dynamics.* New York: Academic Press.

Reice, S. R. 1994. Nonequilibrium determinants of biological community structure. *American Scientist* 82: 424–435.

Chapter 3 ***Recolonization, or How Do All Those Species Fill Up the Gaps?***

If a tree falls in the forest and there is no one there to hear it, does it make a sound?

—Bishop George Berkeley, Ireland, 18th Century

Duke Forest, Chapel Hill, North Carolina, September 1996

In the wake of the destruction wrought by Hurricane Fran, I was anxious to see how the forest would react to this tremendous disturbance. Two weeks after the hurricane, I went into the woods. Downed trees were everywhere. Among the dead and splintered trees, with their branches pointing in weird directions, were the survivors. The small herbs, like wild ginger and spotted wintergreen, were battered but still upright. It was a grotesque fairyland, beautiful and tortured at once. Many trees were killed, yet the herbs were gleaming and reaching toward the light. The aftermath of the hurricane was a mixture of death and rejuvenation. New patterns had emerged. The most striking change was that there was so much more light reaching the forest floor than I remembered. In the misty morning, the shafts of light streaming down seemed semisolid, like columns of pale yellow light. Trails that had been completely shaded were now brightly lit. Downed trees had created gaps in the canopy. Their roots, as they were ripped up and out of the ground, "tilled" the soil. Now, with abundant light, freshly tilled soil, and lots of available nutrients, the earth was well prepared to receive new plants. Which ones would they be?

New Hope Creek flows through the Duke Forest. It is a large stream. Before and during Hurricane Fran, New Hope Creek had received more than one fourth of its typical annual rainfall in less than a week. The signs of a massive flood were all around. Tree limbs and wet mounds of leaves were caught in the top branches of the streamside dogwoods, 9 feet (3 meters) up in the trees. The stream had to rise 18 inches just to leave the banks, then rose another 9 feet up and out onto the floodplain. The flood had pitched big logs 30 feet into the forest. Brand new logjams, 15 feet high, clogged the stream channel. The day after Fran, New Hope Creek was raging at more than 9 feet per second. Its swirling floodwaters were the color of coffee with cream, due to the sand and silt carried along. Like a liquid abrasive cleanser, the water smashed into the rocks with tremendous force. The rock surfaces, which had been covered with algae, mosses, and aquatic insects before the flood, were now scoured clean: sediments were washed out, tumbled, and redeposited farther downstream, sorted out by the flood. Fresh riffles were back in place. The logjams and downed trees created new habitats on the creek bed, and the stream was ready to receive its new colonists. Who was going to arrive?

Disturbances Create Openings

While philosophers—like Bishop Berkeley—puzzle over imponderables such as whether sounds exist independent of ears to hear them (the bishop's answer was "God hears the sound"), ecologists ask a different question. As ecologists, we ask, "If a tree falls in the forest, does it leave a gap?" We answer this question with an emphatic "Yes!" Disturbances create gaps in the living matrix of the community. Recall that it is the process of recolonization that holds the key to increasing biodiversity following disturbance. So, we need to understand how all those spaces get filled back up. To begin to answer this fundamental question, we first have to determine the magnitude of the disturbance. How big was the disturbance? How strong? How severe?

I have shown that the first action of a disturbance is to remove organisms. The more intense the disturbance, the more organisms are removed (fig. 3.1). We can measure the force, strength, or power of a disturbance in many ways. For example, we classify hurricanes

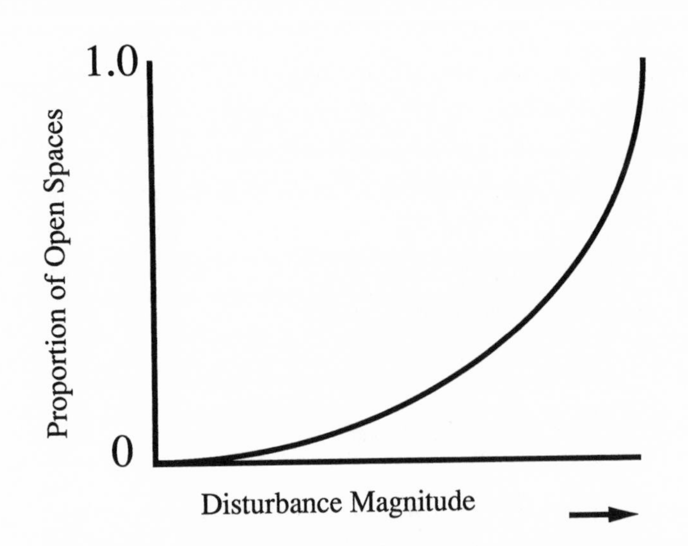

Figure 3.1 The proportion of organisms removed from a community increases with disturbance magnitude or intensity.

by their wind speed. A modest category 1 hurricane has maximum sustained winds of at least 74 mph (or 150 km/hr). Category 5 hurricanes, the strongest ones, have maximum sustained winds of at least 155 mph (250 km/hr). Hurricane Fran was a category 3 hurricane. The extent of a disturbance is its size, or the area impacted. For a hurricane, it is the area with winds greater than 75 mph. For a fire, the magnitude would be the maximum temperature and its velocity. There are "cool" fires (relatively speaking) which may burn the dry, dead wood but not the live trees. Very hot fires can burn down the entire forest and even burn the organic humus right out of the soil. To measure the magnitude of a flood, we would include the total discharge (the amount of water in cubic meters, flowing past a point in the river each second), the depth and velocity of the water, and the area flooded.

In each kind of disturbance, determining its magnitude or strength is a complex combination of the force and its extent. Recall that the force of the disturbance varies spatially, so the strength of the disturbance changes across its extent. We often hear of the "calm in the eye of the storm." In the eye of a hurricane, the winds are nearly still, while at its edge the winds are ferocious. When Hurricane Fran passed over our house, the noise of the wind was terrifying. Then, around 1:30 A.M., it got eerily quiet. Half an hour later, it started to rage and roar all over again.

So, the magnitude of a disturbance is variable in space and time. The bottom line is that the impact, in terms of the proportion of individuals and populations removed by the disturbance, is also spatially and temporally variable. Some gaps are bigger and more open than others. You can think of a gap as a tear in the fabric of the community. A widely held belief among biologists is that "Nature abhors a vacuum!" Ecologists know that if unexploited resources are available, some species will move in to take advantage of them. With this perspective, the dynamics of natural communities are colonization, followed by disturbance, followed by recolonization, followed by disturbance again, ad infinitum. This is a rule of change by which all communities and ecosystems are governed. By this principle, all communities are repeatedly going through succession. So, if a tree falls in the forest (regardless of whether anyone is listening), it will create a gap and the process of recolonization will begin.

Recolonization Brings New Species

Potential colonists are all around us. This is because a fundamental requirement for a species' success is dispersal. Dispersal is an essential part of the life of every species. They are on the move, seeking places to grow and reproduce. Familiar examples of dispersal include dandelion seeds blown off and carried by the wind; oaks producing acorns that squirrels will carry off and bury (thus planting them); young male African lions being driven away from home by their fathers; young birds being kicked out of the nest; and young men and women being sent off to work or college. If all family members were to stay in one place, then the family size would increase and all their resources would get used up. Organisms must therefore disperse and, consequently, potential colonists are always on the move. The interaction between disturbance and spatial heterogeneity creates opportunities for recolonization. The random settlement of new colonists means there will be an increase in diversity following a disturbance. The ongoing process of population destruction followed by recolonization leads to high biological diversity in communities.

How Do Species Recolonize Disturbed Places?

Species can repopulate an open area in only three ways: (1) regrowth of surviving individuals, (2) migration from adjacent patches, and (3) recruitment from a source pool. The modes of action of these three mechanisms are very different, as shown below.

Regrowth is repopulation by survivors of the disturbance. Trees can regrow from stump sprouts. Algae can regrow from basal cells, and grasses from their roots. Sexually reproducing animal populations typically do not regenerate whole individuals, but surviving coral individuals can regenerate an entire reef. The nature and amount of regrowth are predictable. Note that regrowth does not mean local sexual reproduction. Surviving sexually reproducing individuals can surely breed, but this is long-term process; it takes from weeks to many months to repopulate a patch. Regrowth here refers to asexual reproduction (as in corals) or regeneration from surviving tissues (as in stump sprouting by chestnut trees). It is determined by the kinds of species that survived the disturbance, their numbers, and their growth rates. The key feature of regrowth is that only individuals that survive the disturbance can regrow in place. Therefore, regrowth leads to replacement of lost individuals by new individuals of identical species. Only a small fraction of all species can be repopulated in this manner.

Recruitment is largely random. All systems have a wide array of potential colonists. At any given place, at

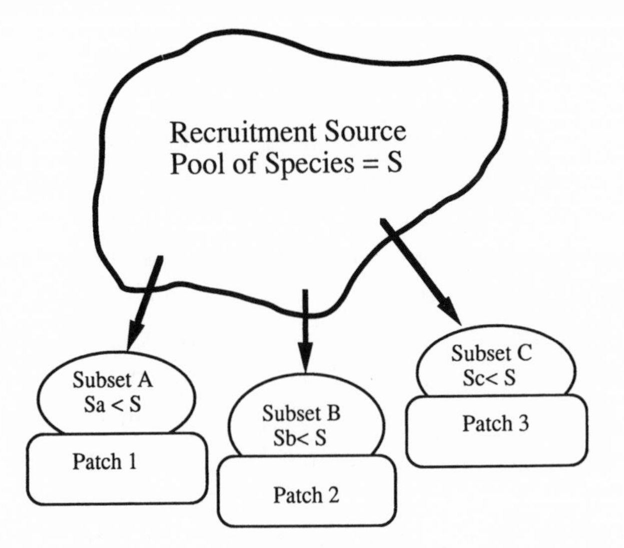

Figure 3.2 Due to random settlement of recruits, even identical open patches will have different arrays of new colonists.

a particular instant, the list of who is available to colonize is due to chance. Available ecological space at any point in time is colonized by a random sample of the pool of potential colonists, that is, the ones that were present at the instant the space became available (fig. 3.2). Imagine that a raccoon, digging for mussels, disturbs a patch of sand in a stream. This local disturbance dislodges or even kills many aquatic insects living in the sand, creating ecological space in the sand. Who will colonize that patch of sand? The next insects, drifting along with the current, reach the "open" sand patch. If the animals that reach it first can live in sand, they will settle there. If they need a different substrate, they continue drifting and keep looking.

The open patch is like a vacant apartment in New York City. If you are apartment hunting, and an apartment comes on the market, and you just happen to be there, and it suits you, you can take it and live there. But if you need a two-bedroom apartment, and the available one has only one bedroom, then you will pass it by. In a stream, it's much the same. The insects drifting by are the potential tenants, and the patch is the home for rent. If it's right for them, they will take it. Animals can be choosy too. There are many individuals in nature—like the Cotton Boll Weevil of the folk song—who are "just lookin' for a home," ready to occupy and thereby recolonize ecological space that was opened up by a disturbance. The recruitment process is variable in space and time and is patch-specific. Thus, different patch types that are adjacent to one another will have unique assemblages of newly recruited species. Even more important in the maintenance of biodiversity is the fact that two patches of the same type (even "identical" patches) opened on the same day will be colonized by different arrays of new recruits. The patchiness of disturbed environments leads to high biodiversity.

Migration is far more certain than recruitment. In contrast to recruitment, migration is colonization from adjacent undisturbed patches. Migration between patches is determined by the species that are available in adjacent patches, while recruitment sources are more remote. Migration is a local phenomenon. The potential new immigrants come from a very short list when compared to the range of species available for recruitment, limited to just those few species that are "next-door neighbors." For example, in a stream, mayflies from a sheltered patch of cobbles may crawl into a flood-scoured patch of cobbles. Similarly, soil arthropods may migrate into burned forest plots from unburned forest plots. Animals move in, vines extend into the newly opened space, and surrounding branches of trees grow toward the light of a new clearing in the forest. Migration occurs from the edges of a disturbance.

A rarely considered question is this: "Why should an individual migrate?" Usually, individuals will migrate only if they are crowded. If all your needs are being met, there is no need to move. Migration will occur if significant crowding exists in a patch, and it may also occur if the conditions in the patch change, making it no longer suitable (for instance, if a predator arrives and drives the current inhabitant out). The principle that "the grass is always greener on the other side of the hill" makes sense for people, who have elaborate cognitive powers. However, it's hard to imagine the individual animal or plant speculating on the availability of distant resources and then acting on that idea. The growth of grass into a newly available patch (with its abundant light and nutrients) occurs because there is nowhere else to grow. All the ecological space where it currently lives is taken. If a plant is not pressed to move, it will stay put. (This is one of the great advantages plant ecologists have over animal ecologists. If you are studying the growth of a tree, and you leave it for a few weeks or months, the odds are that it will still be there when you

TABLE 3.1

Relative Importance of Recolonization Mechanisms of Dominant Organisms in Various Ecosystems

Ecosystem	*Migration*	*Recruitment*	*Regrowth*
TERRESTRIAL			
Deciduous forests	XX	XX	X
Coniferous forests	XX	XX	*
Eucalyptus forests	*	XX	XX
Grasslands	*	XX	XX
Chaparral	*	X	X
Pocosins	X	XX	XXX
MARINE			
Rocky intertidal	X	XXX	X
Coral reefs	XX	X	XX
Mud flats	XX	XX	*
Ocean floor	X	XX	*
FRESHWATER			
Rivers and streams	XX	XX	X
Permanent lakes, ponds	X	X	X
Temporary ponds	X	X	X

Key to degree of importance of recolonization mechanisms: * = unimportant, X = moderate, XX = very important, XXX = extremely important.

come back. Animals generally are not so sedentary or cooperative.) Knowing exactly which individuals will migrate and which will not is an important unsolved problem in behavior and ecology.

The relative importance of recruitment, migration, and regrowth varies widely among systems and species. Table 3.1. shows that some systems are dominated by recruitment (the rocky intertidal ecosystem), with limited roles for migration and regrowth, while others (temperate forest tree communities, for example) have important roles for regrowth and recruitment while migration is less important. The survivors of the disturbance and the array of new colonists determine the reconstituted community structure. High species diversity results from this patch-to-patch variability in disturbance effects and recolonization processes. The new community structure is the result of the newly reconstructed habitat and the arrival of colonists into the available habitat space. The dynamics of some representative ecosystems are discussed in detail below, where I will focus on the dominant space holders, that is, the species that take up most of the room (trees in forests and aquatic insects in streams).

Recolonization in Forests

Recruitment

Plants have a special mechanism for repopulating gaps: their seeds. Seed-bearing plants have a big problem. They need to get their seeds well away from the parent plants so that the seedlings don't compete with their parents for the same resources. Plants have therefore evolved many mechanisms for seed dispersal. Some of these are very familiar. As any child who has ever blown the seed head off a dandelion knows, the seeds will travel a great distance. Kids also like to catch the twisting, twirling maple seeds (*Acer* spp) as they float down from a maple tree on a summer day (perhaps you have even peeled one open and stuck it on your nose). Many plants have evolved to exploit animals to solve their dispersal problems. You may have pulled burrs from your dog's coat or picked the beggar ticks off your jeans after a walk in the woods, thereby spreading the seeds of *Bidens* spp.

Fruit is the best seed-dispersal trick of all. Fruit-bearing plants produce seeds wrapped in a sweet, edible pulp. Animals eat the fruit, and the seeds pass undisturbed through their guts. Sometimes the digestive enzymes can initiate a seed's germination. A bird eats a cherry and "poops" it out, now covered with fertilizer. That's "airmail special delivery"! You eat an apple and toss away the core, loaded with seeds. You are continuing the work of Johnny Appleseed, dispersing apple trees. Squirrels collect acorns (oak seeds) and bury them. However, because squirrels don't have very good memories, they often forget where the buried seeds are and never come back to eat the ones they buried, or planted. Thus, the oaks can germinate in the new homes provided by the absentminded squirrels. All these dispersal methods take the seeds into a new habitat and away from the parent, to grow into new plants.

How does all this relate to recolonization of disturbed forests? When seeds are dispersed, not all of them germinate at once. Some seeds are viable (i.e., capable of developing into a young plant) for only a very short time. Willow (*Salix* spp) seeds, for example, are viable for just a few days. If they don't land in a suitable habitat, they will simply die. Features such as soil moisture and light level determine if a habitat is appropriate for a species. Other seeds can remain viable for many years: Danish archeologists were able to germinate 1,700-year-old seeds of Lambs Quarters (*Ameranth* sp., sometimes called pigweed), and Oriental Lotus from China still had some seed viability after 3,000 years! Fireweed, *Erechtites hieracifolia* (fig. 2.6), could also be called Hurricane Weed, according to Peter White, director of the North Carolina Botanical Garden. Its seeds remain viable in soil for twenty or more years, where they are stored as part of the "seed bank." Virtually every potential resident plant species has seeds in the soil wherever it can grow, waiting for the proper conditions to germinate. Densities of seeds in the seed bank have been estimated at over 100,000 dormant seeds per square meter. A fire or blow-down produces large, open gaps in the canopy of the forest, now brightly lit with sunshine. So, *Erechtites* seeds were just waiting (patiently) in the soil when Hurricane Fran gave it the golden opportunity to germinate and make the immense patches of Fireweed that we saw the next summer. This germination of new plants in disturbed soils can be thought of as delayed recruitment, since the seed source originally came from outside the disturbed area, then lay dormant until the disturbance created the opportunity for it to germinate. Animals also recolonize forest gaps by recruitment. Birds fly into disturbed patches from remote sites. Similarly, rodents (mice, voles, and squirrels) can travel a long way to colonize a gap. Butterflies, bees, moths, and mosquitoes will fly into and settle a vacated patch.

Migration in Forests

Migration is easiest to envision by considering animals. If a tornado hit a forest patch and falling trees killed all the mice in the area, then other mice will migrate into the open patch from the overcrowded areas surrounding it. This mechanism is important for everything from beetles and soil arthropods to earthworms and mammals. Most plants migrate very slowly. However, some plants are highly aggressive and spread rapidly into gaps created by a disturbance by way of vines, runners and rhizomes. The colonization comes from healthy individuals on the edge that grow and spread into the gap.

Consider *Ajuga* spp (commonly called Bugleweed), a low green to purple, herbaceous ground-cover plant. *Ajuga* spreads by rhizomes, which are underground lateral stems that periodically send out roots at nodes to start new plants. I intentionally planted twelve 2-inch diameter plants in a small, 9 by 12 foot, garden five years ago. The original plants covered less than 0.1 percent of the area. Now the garden is solid *Ajuga*, which spreads very aggressively. If I wanted to get rid of it now, I doubt that I could. A common aggressive vine is Japanese Honeysuckle (*Lonicera japonica*), a highly invasive plant that spreads very rapidly. Vines run along the ground and put out leaves and sprouts along their length, but not at discrete nodes. A ubiquitous and well-known vine is Poison Ivy (*Toxicodendron radicans*). Its ability to spread was even memorialized in song by the Coasters in the 1950s: "At night when you're a-sleepin', poison ivy comes a-creepin' around." Strawberry (*Fragaria virginiana*) spreads with above-ground runners (it's my daughter's favorite runner). Runners are thin stems that creep along the ground and put out roots at nodes along its length, spreading the plant. Note that plants that migrate using runners or rhizomes are genetically identical to the original plant. The new plants are really part of one clonal individual.

A common tree that spreads (migrates) by rhizomes is the Aspen (*Populus* spp). It is no coincidence that Aspen is one of the most geographically widespread trees on the planet. The evidence of the existence of clonal individuals is apparent in the Rocky Mountains. One whole stand of Quaking Aspen (*Populus tremuloides*) will turn golden on a single autumn day. A neighboring stand may wait a week to change and then change all at once. These are two genetically distinct clonal stands of identical trees. Sweetgum (*Liquidambar styraciflua*) is an eastern tree that also spreads by rhizomes.

The competitive pressure (in the form of shading or root competition) from neighboring trees reshapes most trees in a forest as they compete for light, water, or soil nutrients. As a result, the idealized crown shape of the tree species is only manifest when the tree is growing out in the open, as in a meadow. Otherwise, the trees must grow when and where they can, with distorted crowns. When a disturbance opens up a gap, the trees along the margins of the gap can expand and grow into it. This too is a form of plant migration. The most extreme example of this is Sourwood (*Oxydendron arboreum*), which grows directly toward the light. It will simply grow into the gap by extending its branches (or even its trunk) into the available space. There are Sourwoods with trunks that are nearly parallel to the ground as a result of growing toward a light gap.

Regrowth in Forests

Regrowth is the growth of an individual organism from a remnant part. In plants, it occurs most commonly from roots. This capacity to regrow is well known to those who have mowed their lawn in the hope of eliminating dandelions (*Taraxacum officianale*). Within a week or two, the dandelions are back, full strength: they simply regrow in place. You can get rid of dandelions only by digging out the entire root system. (Sorry, but pulling them out simply doesn't work.)

Oak seedlings in the Piedmont forests of North Carolina that appear very small are, upon close examination, often very old. Dr. Lopez-Mata has concluded that these small old plants are the result of repeated browsing by White-tailed Deer. The deer chop off the oak before it gets large and tough, and the oak regrows in place, again and again and again, keeping the tree small. This is nature's way of creating a bonsai plant. Bonsai, the

Japanese art of creating miniature trees and forests, works on the same principle, where repeated pruning keeps the trees very small. This is simply regrowth turned into an art form. Regrowth is an effective way of replacing an individual whose above-ground portion has been removed by a disturbance. Another type of regrowth is through sprouting from a tree stump or downed trunk, when the root system of the tree is still intact. Trees, which are renowned for their ability to produce viable new individuals from stump sprouting, include the Oaks (*Quercus* spp) and Red Maple (*Acer rubrum*). There are no examples of terrestrial animals that can regrow in place!

Thus, fires and tornadoes frequently disturb forests. When gaps are created by the disturbances, the gaps are recolonized by all three methods. Each is important, but migration and recruitment are far more common and significant than regrowth.

Recolonization in Streams

Recruitment

There are two prominent sources of insect recruits in streams. The first is the deposition of the eggs of stream organisms into the stream. Aquatic insects go through four developmental stages. They start as eggs, grow and mature as larvae, then change into pupae (the pupal stage is a step toward full adult morphology, or body form). They then emerge from their larval home in the stream or pond as winged adults. Now they can fly. They mate, then lay their eggs back in the water as a new, external supply of new individuals. These are colonists of the stream. When they settle into a disturbed patch, they become the recruits. They lay the eggs in a patch of water, based only on the appearance of the water. We know of no way that the adult female can judge what the local abundance of insects in the stream is, so her choice is independent of the aquatic community and its density. Looking up from the denuded, disturbed patch of the stream bottom, the arrival of a batch of eggs appears to be completely random.

The second and more important mechanism is stream drift, which looks random, too. Every night, more than one percent of the total stream fauna floats downstream with the current. This includes everything from stoneflies and mayflies to worms and snails. Stream organisms release their grip on the substratum and are carried passively by the current. These drifting organisms eventually settle in a new place. If the place they stop suits them, they stay there. If not, they release themselves from the substrate and drift on downstream until they reach a suitable patch. A vast amount of stream ecology research has been devoted to drift to answer questions such as: Why do the animals let go of the substrate in the first place? Is drift active or passive? (Probably both at different times.) Is it due to predation pressure, competition, or crowding? (Probably yes in all three cases.) Aquatic plants can also drift. Drift is the principal mechanism for repopulating disturbed areas in streams.

An interesting paradox arises. If all organisms eventually drift downstream, then the upstream reaches should be empty and the downstream reaches jammed. The Colonization Hypothesis explains the paradox: if the insect eggs are being deposited upstream, and then insects drift downstream, the problem is resolved. Anne Hershey of the University of North Carolina at Greensboro has new evidence that when chironomids (midge larvae) form mating swarms in the winged adult stage, they do indeed fly upstream. She used genetic markers to prove that the new chironomids upstream were related to the downstream populations.

Migration in Streams

Migration is the movement from the undisturbed community around the edges of a disturbed patch. Repopulation of upstream reaches by insects swimming or crawling upstream against the current is a form of migration. Others move laterally across the stream bed or crawl up from beneath the surface sediments. Different stream disturbances (floods or droughts, a toxic spill, or cattle trampling the stream bed) have different effects on different patches. Some are highly disturbed and open to recolonization, while others remain virtually undisturbed. Animals in the undisturbed patches (if crowded) will move into an adjacent (open) disturbed patch, where space and other resources are available. I (and others) have done many experiments that demonstrate the reality of this scheme. If I introduce a block of open substrate (for example, a basket of gravel or a bundle of dead leaves—a leafpack) into the stream and I return and sample it after a day or a week, I will find many new colonists present. Some have crawled in (migrants) and some have drifted in (recruits). The more mobile the organism, the more likely it is to be a colonist. Blackflies and mayflies are often very early colonists, while more sedentary animals such as sponges and case-building caddisflies come in much later. Plants migrate, too, but very slowly. Some algae, aquatic mosses, and macrophytes move into denuded stream patches by the same mechanisms as forest plants.

Regrowth in Streams

Very few stream organisms can regrow in place. Algae can regrow from basal cells that are still attached to rocks after a flood. Animals such as hydra (Coelenterata) can regrow if a portion of their body is still attached to rocks. These organisms are not very common and are present predominantly in slower streams. Sponges can also regrow in place. These are often seen as a thin covering on rock surfaces in riffle zones (fast water). With these few exceptions, regrowth is rare in streams (table 3.1). So, we can see that drift and egg laying (both forms of recruitment) are very important, contributing to increased stream biodiversity after disturbance. In addition, upstream and lateral migrations are also key sources of new colonists for open stream patches.

Disturbance, Recolonization, and the Maintenance of Biodiversity

Disturbance per se does not determine diversity. Disturbance creates opportunities for colonization of new individuals and new species into the vacated space. If regrowth dominates the recolonization process, then individuals that colonize a disturbed patch will be the same as those removed by the disturbance. Similarly, if immigrants are identical to the former residents, there will be no change in community structure and no increase in species richness. For diversity to increase, the supply of new colonists to a disturbed, open patch must be different from the residents in the surrounding patches. The random draw of recruits from the broad species pool assures variation, and hence the maintenance of high diversity. Recolonization mechanisms of migration and regrowth stabilize biodiversity, but recruitment enhances it. Following a disturbance, all systems have an array of mechanisms for recolonization. The relative contributions of these three mechanisms clearly vary widely among communities (table 3.1). Recolonization by "random" recruitment provides the widest array of colonizing organisms that produces high biodiversity. Recruitment is the key to disturbance-induced biodiversity.

How does the importance of the contributions of the three recolonization mechanisms change as the size of a disturbance increases? Regrowth doesn't change at all. Regrowth is a function of the disturbance intensity more than its size. Powerful disturbances (e.g., hot fires, volcanic eruptions) can kill everything. Mild disturbances leave more individuals present, some with the ability to regrow. Differences in disturbance intensity occur whether the disturbance is large or small. These differences are not dependent on the extent of the disturbance. A boulder carried by a flooding river can act like a battering ram. It can smash every caddisfly on a rock in the stream, or it can bump the rock and hardly kill anything. Hurricanes can be category 1, with a large extent and inflicting relatively little damage, or category 5, with a small but devastating area of impact.

However, there is a significant shift in the relative contribution of migration and recruitment as the size of a disturbance increases. Big patches have proportionately less edge than small patches, with an important ecological consequence. Recruitment is random, so it is essentially equal at all points in space; the larger the disturbed area, the greater the total number of recruits. However, migration occurs across the edge (perimeter). The amount of edge relative to the total disturbed area decreases as the disturbed area increases. So, the proportion of new colonists provided by migration is directly proportional to the ratio of the perimeter to the area disturbed. Therefore, the proportional contribution from migration actually decreases as the disturbed area increases, and the proportion of colonists from recruitment increases with disturbed area (fig. 3.3). This is true for any ecosystem. Recall that the recruits are a random sample of the total species pool available at that instant

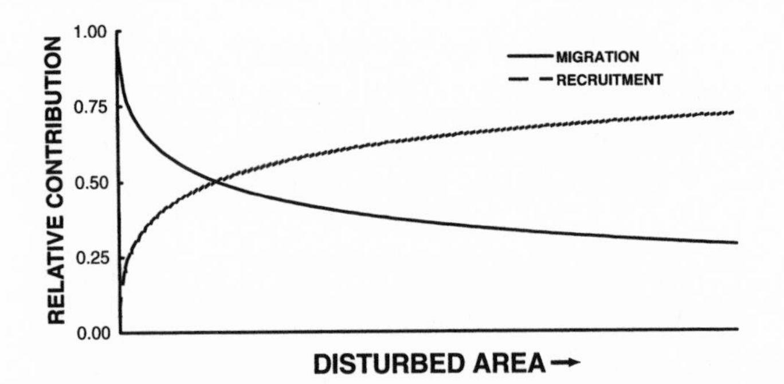

Figure 3.3 As the size of a disturbed area increases, the proportion of colonists arriving via migration and recruitment varies inversely. In large disturbances, recruitment is the main source of new colonists; in small disturbances, migration is the main source.

in that particular place. The migrants are a subset only of the immediate neighbors. So, recruitment is more likely than migration to bring new (unrepresented) species into the disturbed area. The greater the contribution of recruitment to the recolonization mix, the greater chance there is of increasing species richness for the entire system.

Counter to the popular notion, fostered by the media, that disturbances are only disasters and tragedies, the reality is quite the opposite. Disturbances permit and support the maintenance of high biodiversity in nature. We need to learn to embrace them rather than fear them. Disturbance is a two-edged sword. As it cuts down some species, it creates opportunities for others. Disturbance is the key to the survival and coexistence of many species in the face of their more powerful competitors. Disturbances are vital to the preservation of biodiversity, which benefits the ecosystem and us. If every cloud has a silver lining, this silver lining is biodiversity and must be appreciated and valued.

Further Reading

F. S. Chapin, B. H. Walker, R. J. Hobbs, D. U. Hooper, J. H. Lawton, O. E. Sala, and D. H. Tilman. 1997. Biotic control over the functioning of ecosystems. *Science* 277: 500–504.

Darwin, Charles. 1988 [1859]. *On the Origin of Species.* New York: New York University Press.

Fowler, C., and P. Mooney. 1990. *Shattering Food, Politics and the Loss of Genetic Diversity.* Tucson: University of Arizona Press.

Wilcove, D. S., D. Rothstein, J. A. Dubow, A. Phillips, and E. Losos. 1998. Quantifying threats to imperiled species. *BioScience* 48 (8): 607–615

Chapter 4 **Disturbance Ecology and Fire Management: "Let It Burn!"**

The world, an entity out of everything, was created by neither gods nor men, but was, is and will be eternally living fire, regularly becoming ignited and regularly becoming extinguished.

—Heraclitus (c. 535–c. 475 B.C.), Greek philosopher, *The Cosmic Fragments*, no. 20 (c. 480 B.C.)

I am the Lorax. I speak for the trees.

—Dr. Seuss, *The Lorax* (1971)

Laguna Hills, California, September 1995 (or 1996, 1997, 1998, 1999, 2000, 2001, . . .)

Chaparral is the Spanish word for Mediterranean shrubland vegetation. This is arid land, inhospitable to most organisms, but harboring a unique community of animals and plants. Low shrubs are clumped together or very sparse, depending on the availability of water. These plants are squat, rarely more than 2 feet tall, with very thick waxy leaves. The low profile and the protective wax covering are adaptations to survive in dry conditions. The vegetation's color is muted, ranging from dusty pale green to nearly gray.

For three weeks, the chaparral had been parched by hot, dry weather with strong winds. No one really knows how the fire started. Was it a tossed cigarette? An unattended campfire? Lightning? Whatever the source of the spark, the chaparral was quickly ablaze. Driven by the Santa Ana winds, the scene was soon a conflagration. As the waxes in the leaves melted, they vaporized and burst into flame. Flames reached 0 to 30 feet (7–10 m) into the sky. Soon the whole hill slope was on fire and the flames raced and darted about, pushed by the wind. Hill after hill exploded into flame. The chaparral was burning . . . again.

Is this a disaster? Not at all. As explained in chapter 2, chaparral is built to burn. The seeds of these plants will not germinate unless a hot fire heats them, so chaparral requires fire for its very survival. Fire is completely normal in this ecosystem. Why, then, do our national TV news anchors call this a tragedy? Perhaps it is because there are people's homes in those hills, big, elegant homes, $500,000 to $1 million homes. They are in the path of the fires. They were not built to burn, as the chaparral is, but burn they will.

Perhaps, in a certain sense, this is a tragedy. In Greek tragedy, the terrible event befalls the hero because of a flaw in his character, or because of a fatal error in judgment. The key element in tragedy is that the terrible thing could have been avoided but for the "tragic flaw" in the protagonist. It is the action of the hero that turns him into a victim. If there is tragedy in this typical Southern California fire, it is that the wealthy homeowners did not know or want to believe that they were building on a tinderbox. We have known for thirty years that chaparral is a fire-dependent community. Ecologists know that the chaparral will burn about once every fifteen years in order to survive as a community. This fire, in this exact place, at this precise moment, could not be predicted. However, the reality of a fire occurring here, sometime soon, was a dead certainty. What is the tragedy, then? The tragedy is that the homeowners, not knowing about the fire dependence of the chaparral ecosystem, or just choosing not to know, have put themselves in harm's way. As in a Greek tragedy, through their actions, the homeowners may lose their homes or even their lives.

"Fire Resistant": The History of U.S. National Parks and Forests Land-Management Policy

In the United States, we have long been "fire resistant." The policy of the land managers of the United States since 1940 has been to suppress all fires in the national parks and national forests. One national forest supervisor put it this way: "Any fire spotted in the morning will be extinguished by sundown." Smokey Bear, the emblematic cartoon spokesman for the U.S. Forest Service, with his stern yet friendly face and deep voice, intones, "Only *you* can prevent forest fires." Smokey's image has been hammered into all of us, in billboards, television, and print advertising. Smokey's message is clear: "If you love forests, you must save them from fires. Fires destroy forests." The intention has been to "protect " our forests and wildlands. Perhaps the most searing image we grew up with was the forest fire scene in the Walt Disney movie, *Bambi*, which was terrifying to both children and adults. Animals were burned alive. The fire killed Bambi's mother. The result of the fire was death and devastation. Fire was clearly the enemy of nature. However popular these messages are, they are also false.

This policy of fire suppression is based on the notion that forests should remain undisturbed to maximize their beauty and economic value. It is part of the whole concept that disturbances are bad for nature, so that the burning of a forest is "a tragic loss." This has been translated into the national policy that we should not allow forests to burn because "fire is bad for forests." This policy has backfired. The policy of fire suppression is an expression of the equilibrium hypothesis of community structure. Let's examine the ecology of fire and fire suppression in light of the contemporary theory of disturbance ecology.

Does Fire Damage Forests? Does Fire Suppression Help Them?

These questions are not so simple or easy to answer. When we hear on the news that "100,000 acres of forest were destroyed," that statement is fundamentally untrue because the fire kills some trees and favors others. On the one hand, fires will surely kill some trees. Hot ground fires can roast the fine roots, denying the tree the ability to absorb and transport essential water and nutrients to the leaves. Crown fires that consume large amounts of branches and foliage prevent the tree from producing sufficient food (through photosynthesis) to survive. Fires that are hot enough for long enough can burn entire trees. An uncontestable aspect of fire is that it kills trees. On the other hand, fire plays a variety of roles that can be beneficial to forests, in particular, to the maintenance of biodiversity. Even as some trees are killed, others survive. Let's first consider what happens when fire is suppressed.

Fire is a disturbance in forests. As any disturbance, it is unpredictable. It kills resident individuals. It opens up ecological living space for recolonization. In a forest where fire has been suppressed for many years (say fifty or more), the fire-intolerant trees grow unchecked: they get large, the canopy closes, and light reaching the forest floor is sharply reduced. Trees that are dependent on high light intensity are stunted, or they may be prevented from growing at all, and as they get weaker, these trees eventually whither and die. The understory trees are limited to shade-tolerant species. We will ultimately lose all the fire-dependent species. As in the Intermediate Disturbance model, the competitively dominant trees displace the weaker species and overall biodiversity is reduced. What has happened is that, in the absence of fire, the growth of the dominant trees has rendered the habitat unsuitable for a large portion of the potential forest species. This can so change the species composition that the character of the landscape may be changed beyond recognition. In the western part of Great Smoky Mountains National Park, fire was a common visitor. The return interval of fire, prior to 1940, has been estimated at only twelve and a half years. This means that, on average, a given stand of trees burned approximately every dozen years. This number was derived from the appearance of black bands of charcoal in tree cores. Since fire suppression became national land management policy, the return interval has now been estimated to be five hundred years! Now, fires are rare and far apart and burn only small areas. The biggest change in the Great Smoky Mountains forests is the tremendous increase of Red Maple, a fire-intolerant species, to the detriment of the fire-tolerant pines. Similarly, with fire suppression, a chaparral ecosystem becomes scrub oak woodland. A longleaf pine forest without fire becomes an oak-hickory forest. The Florida Everglades, without fire, will eventually be transformed into oak woodlots.

All fires require an ignition source, tinder, fuel, and air. The source of ignition can be natural, such as a lightning strike, or anthropogenic (human-caused), like a carelessly tossed cigarette or match. As more and more people seek a wilderness experience or outdoor recreation, the potential for anthropogenic ignition sources increases. The U.S. Forest Service has recognized this issue yet still addresses it through the admonitions of Smokey Bear. However, a spark alone will not cause a fire unless there is flammable material—tinder—around for the spark to fall onto and ignite. Such tinder includes dry leaves, twigs, and branches. With fire suppression, the tinder and fuel loads build up. Fire suppression results in bigger, hotter, more devastating fires that can change the essential nature of the forest stand or even really destroy it.

How Does Fire Encourage Plant Growth?

Fire has such a bad reputation that we commonly forget that fire can actually encourage and stimulate plant growth. For many plants, fire is necessary for their survival. Fire creates the environment that many plants need to reproduce and, indeed, to live. Consider the bed of pine straw in a pine forest or the thick mulch of leaves on the floor of a deciduous forest in autumn. These needles, leaves, and dead tree trunks and branches are the repository of nutrients, which the trees have taken from the soil and used in growth during the spring and summer growing season. As the plants photosynthesize and grow, those nutrients are bound into complex organic compounds (i.e., proteins, carbohydrates, and fats). Although plants have no cholesterol, which is a uniquely animal fat, they certainly do produce fats, from corn and safflower oil to the waxy coating on magnolia leaves. If these organic compounds were not broken down, then all the nutrients would get bound up in living and dead plant tissues and stay there. Life, as we know it, would cease. So, decomposition is a vital process in forests, indeed in all ecosystems.

However, there is a problem. Decomposition is typically very slow. A dead tree trunk on the forest floor can persist for up to fifty years, even longer in a dry climate. The dead leaves and pine needles build up a rich organic layer until decomposition converts them into humus. Between fires, the organic matter builds up and the growth of many species declines. Fire is an important way of speeding up the pace of recycling and encouraging plant growth. The fires produce an ash rich in plant nutrients. In the dead leaves and branches, the organic nitrogen (N) and phosphorus (P) is bound up in complex molecules and therefore is unavailable to living plants. Fire, by burning the organic matter, is oxidizing the organic nitrogen N and P. In the ash (the inorganic remnants of the fire) the N and P are now in the oxidized form of nitrates (NO_3^{-1}) and phosphates (PO_4^{-3}). These nutrients can now be readily absorbed and utilized by the surviving plants. The combustion of the trees also opens up the canopy, letting in more light. Fire thus acts to increase the availability of light and nutrients, which, with water, are the requirements for all plant growth.

Why Fire Suppression Fuels Bigger Fires

Now we can see that fire can be either terribly destructive or actually beneficial. This distinction turns on the intensity and magnitude of the fire. Fire susceptibility in forested ecosystems depends on, mainly, fuel quantity and fuel quality. The risk of fire is not uniform across the forest, so the intensity of the fire, once it starts, will also be variable in space and time. The variation in fire susceptibility across the landscape was shown in figure 2.4. If there is little fuel available, the fire will remain a relatively cool ground fire and burn itself out. Under a fire suppression regime, the fuel load builds up because there is more tinder and fuel wood on the forest floor, and there are more standing dead trees. Under fire suppression, the susceptibility to fire increases everywhere. Under long-term fire suppression, the fuel load builds up and the fire, when it eventually starts, will be hotter, more intense, and more destructive than under a natural fire regime.

Why was the 1988 Yellowstone fire so destructive? Yellowstone National Park was created as our first national park in 1872. Fire suppression has been the U.S. Park Service Policy since 1940. Consequently, for forty-eight years, dead wood and tinder accumulated, just waiting for a spark. The hot dry conditions made the forest ready to burn, and the tremendous supply of fuel guaranteed that the fire would be a big one. The strong winds spread the fire, and it quickly became one our largest fires ever. The Yellowstone fire became a crown fire, with 1,585,000 acres of forest in the Greater Yellowstone area burned, almost 800,000 acres inside the national park. A tremendous amount of wood was consumed. Yet for all the destruction, the forest was not destroyed. In fact, the next spring brought a veritable explosion of new growth: wild-

Figure 4.1 Ten years after the fires of 1988 in Yellowstone National Park, the bed of Lodgepole Pine seedlings beneath the burned-out trunks is remarkably dense.

flowers bloomed in profusion, and there was a flush of new Lodgepole Pine seedlings. In an estimate by Margaret Fuller, counts in 1989 ranged up to 316,000 seedlings per acre in burned areas. The profusion of Lodgepole Pine seedlings beneath the burned-out trunks is remarkably dense ten years after the fire (fig. 4.1). Species that had not been seen for decades were suddenly thriving. Aspens were stimulated by the fire to sprout, as clones, and produced up to one million sprouts per acre. Some of these sprouts could grow to be as much as ten feet tall in only six years.

The forest was not destroyed—it was *rejuvenated*, a nearly perfect word. Rejuvenated means "made young again." The fire removed the old-growth, senescent forest. Niche space was created for colonizing species. The forest was set back to an earlier successional stage; it was reborn. If we had understood the risks of fire suppression and permitted small fires to burn and reduce the fuel load, we would not have faced the horrifying prospect of a mega-crown fire in the jewel of our National Park System.

Disturbance Ecology and Land Management Policy

It is unquestionable that the major aim of the National Park Service and the National Forest Service is to protect and preserve our natural heritage. A problem emerges when we, the people, insist that our national parks and forests perform diverse and often contradictory functions. The National Park Service was created in 1916 by a law called the Organic Act. It states, in part, that the Park Service should *"conserve the scenery in natural and historic objects and the wildlife therein and to provide for the enjoyment of the same in such a manner and by such means as will leave them unimpaired for the enjoyment of future generations."* That is a tall order. We want our national parks to be natural museums, preserving the nation's natural wonders, but we also want them to be available for tourism on a massive scale. Under our policy of "multiple use," we insist that our national forests provide a major source of wood products for the timber industry, but we also want them to remain as pristine wilderness areas for camping and hiking and nature study. These divergent goals give our national forest and park managers an impossible task. How can they possibly achieve all these contradictory goals? Visitation to the national parks is at an all-time high. Over ten million visitors came to the Great Smoky Mountains National Park in 1999. How does one open a park to tourism and still "conserve the scenery in natural and historic objects and the wildlife therein"?

We cannot blame the managers, because they are responding to our will as expressed through legislation passed by our elected Congress. A big part of the problem is that we, the people, don't understand that wilderness preservation is incompatible with fire suppression. When we deny our ecosystems their reinvigorating, rejuvenating fires, we allow them to grow old, to become senescent. When fire-dependent ecosystems mature, in the absence of fire, their entire character changes. The fire-maintained system is replaced by something entirely different. In essence, we lose that unique system, together with all of its biodiversity. Understanding our mistakes is the first step to correcting them. The further challenge is to learn to live with fire.

Examples of Ecological Damage Caused by Fire Suppression

Fire and the Boreal Forest

My first encounter with the ecological benefits of fire came in my first summer of graduate school at Michigan State University. The north woods (the boreal forest) of

Michigan, Wisconsin, and Canada have stands of Jack Pine (*Pinus banksiana*) forests. Mature Jack Pines have thick bark with low flammability so they can survive a fire. Jack Pine is a serotinous species, meaning that its cones will not open to release the seeds until heated by fire to greater than 45°C (113°F). Seeds remain viable in the cones for at least ten years, just waiting for the fire and its heat to allow them to open and disperse. It is a critical adaptation because the Jack Pine seeds will not germinate except on bare mineral soils. Between fires, pine straw and leaves (duff) build up on the forest floor. When a fire occurs and burns away the duff, the mineral soil is exposed. Then the habitat becomes ideal for Jack Pine seed germination, and that's just when the cones open and release the seeds. Furthermore, Jack Pine is intolerant of shade. The fire opens up the canopy, prepares the seed bed, and provides ash for fertilizer. In Jack Pine, evolution has produced a synchronous reproduction cycle in response to the fire. Furthermore, the seeds have short dispersal distances so that the Jack Pine stands persist locally. We can only marvel at the adaptations of the Jack Pine's life history to fire. Could a trained nurseryman do any better than this?

The Kirtland's Warbler (*Dendroica kirtlandii*) lives in the Jack Pines and feeds on its seeds. The suppression of fire is causing the Jack Pines to disappear, since they cannot reproduce without fire. Consequently the Kirtland's Warbler is threatened with extinction. The realization that fire had positive benefits for an endangered species and its ecosystem sparked my interest in disturbance ecology.

Fire and the Longleaf Pine Savanna

In the Sandhills region of southeastern North Carolina are the towns of Southern Pines and Pinehurst. The towns were named after a special ecosystem, the Southern Pine Savanna, which is dominated by majestic Longleaf Pines, *Pinus palustris.* Its needles can reach 18 inches and the cones are enormous, often 12 inches long. When I first saw these magnificent cones, I was so charmed that I collected a dozen of them, and one still graces my office. The Longleaf Pine Savanna ecosystem once stretched southward from Virginia to Florida, and west to Texas, covering 92 million acres. Now less than 4 million acres remain. It is a tattered remnant of its former glory and exists in only relatively small, scattered patches. The Longleaf Pine Savanna ecosystem is vanishing. From the 1850s until the 1930s, Longleaf Pines were the major source of turpentine and pine tar. The trees were scarred, in a practice called boxing, and the sap collected. Repeated boxing killed millions of trees, and logging took millions more. Only 3 percent of that ancestral forest remains extant today, with only a fraction of that in old-growth forest. The surviving Longleaf Pine Savannas are falling prey to fire suppression. They are further threatened by development.

Where the natural fire regime has been maintained, this community is one of the most diverse plant commu-

Figure 4.2 Carnivorous plants in the Longleaf Pine Savanna. (a) Pitcher plant. (b) Venus fly trap.

nities in the world. For example, the Green Swamp in southeastern North Carolina has seven species of carnivorous plants (fig. 4.2). The Longleaf Pine Savanna has four subspecies of the fox squirrel (*Sciurus niger*) and is home to the threatened Golden Tortoise (*Gopherus polyphemus*). The Golden Tortoise is a keystone species, in that its presence facilitates the survivorship of some three hundred other species that directly or indirectly benefit from its presence. It digs burrows in the sandy soil, which many other species use for shelter, including gopher frogs and indigo snakes. Old-growth Longleaf Pine is home to the Red Cockaded Woodpecker (*Picoides borealis*), an endangered species that excavates large nesting cavities high in the trunks of large, ancient Longleaf Pines; many other species subsequently use the abandoned nest cavities for their homes. These old-growth trees have survived repeated fires, which have also kept the oaks from encroaching. The Longleaf Pine Savanna thus supports an entire community of animal and plant species, many of which are threatened or endangered species. They are at risk because their homes are being destroyed.

In this community, the dry spring and early summer fires are larger and more intense. They get hot enough to burn oaks and other hardwoods (excluding them from the system), but are tolerated by the Longleaf Pines. Late summer fires are more frequent but less intense. Longleaf Pine, the dominant tree, is well adapted to fire, becoming highly fire resistant by its second year. The new buds are encased in a slow-burning resin, which lowers the internal heat and allows the buds to survive. The Longleaf Pine produces flushes of needles intermittently, so after a fire the trees quickly produce new needles and scarcely miss a beat in their growth. The tree also grows rapidly; the bark gets thicker (and acts as an insulator) and the trees grow tall in a few years (fig. 4.3). They shed their lower branches, making it harder for a fire to reach the crown and kill

Figure 4.3 Longleaf Pine forest in Carteret County, North Carolina, showing the fire-blackened bark from previous ground fires and the dense post fire undergrowth. Many of the young Longleaf Pines are growing rapidly.

the tree. So, this species is vulnerable to fire for only a short period of time.

The adaptation to fire in the Longleaf Pine community is not limited to the pines. Groundcover plants are mostly perennials that resprout from buds on their rhizomes and are fire resistant. Most ground covers flower right after a fire. The dominant grass, *Aristida beyrichiana,* flowers and produces viable seeds only after fires that occur during the growing season. *Pityopsis graminifolia,* the Golden Leaved Aster, also a dominant plant, increases its growth (forming large clones) and flowers after fires early in the growing season. The entire community is adapted to fire.

Furthermore, Longleaf Pines require bare soil for their seeds to germinate and begin to establish new trees. The availability of bare mineral soil is ultimately essential to maintain the existence of the forest. As the needles at the bottom of the pile decompose, a new layer of needles is deposited on top. If decomposition alone were the only way to remove the waxy, slowly decomposing needles from the forest floor, it would essentially never happen. When a rich organic layer of decomposing needles lies on the ground, this not only denies *Pinus palustris* the opportunity to geminate and live, it also favors the germination and early growth of oak trees. The oaks, which will colonize the Sandhills, include Southern Red Oak (*Quercus falcata*), Virginia Live Oak (*Q. virginiana*), Turkey Oak (*Q. laevis*), Post Oak (*Q. stellata*), Laurel Oak (*Q. laurifolia*), and Myrtle Oak (*Q. myrtifolia*). These oaks, once established, will outcompete the pines and replace them. It is like an old established neighborhood in the inner city, which supports a rich vibrant community, being torn down (for progress!) in the form of condominiums for the nouveaux riches. Without fire, the Longleaf Pine Savanna will disappear into a thicket of oaks. The real threat comes not from fire, but from fire suppression. Without fire, we will lose the savanna and all of its unique species. An entire ecosystem is threatened with extinction.

Eucalyptus Forests in Australia

Some trees have a very special adaptation to fire: they decorticate; that is, they shed their bark the way many trees shed their leaves. These species create their own tinder and actually invite fires. Sycamore (*Platanus* spp) is a common north temperate tree that decorticates in the summer, that's why its bark looks so scaly and patchy. What does a tree shedding its bark have to do with fire?

Let's consider the *Eucalyptus* forests of Australia. When you walk in one of these mature forests, the ground is covered, ankle deep, with the long strips of dead, dry, thin bark. The bark is rich in volatile, highly flammable oils (*Eucalyptus* oil, of course). These *Eucalyptus* forests have evolved to encourage fires that scorch but rarely kill the trees. The fire converts the dead bark to nutrient-rich ash, thus fertilizing the trees and helping them grow. The fire also kills seedlings of competing species, so that the forest succession is arrested by fire. This allows the forest to maintain itself as virtually pure *Eucalyptus* stands. The pale gray-green leaves hanging down from the branches give a nearly mystical quality to the forest, enhanced by the gentle crunching of the bark underfoot. *Eucalyptus* species have evolved to exploit fire to perpetuate themselves.

Chaparral

In the opening passage of this chapter, I suggested that chaparral communities "invite" frequent fires. Such systems are called *fire-dependent ecosystems* because their health depends on frequent fire. The chaparral vegetation is adapted to fire and has even evolved to facilitate ignition. The leaves of Snowbush (*Ceanothus*), a common chaparral plant, are swaddled in a flammable resin.

In order to germinate, its seeds must be heated to 113°F (45°C) for at least eight minutes. The seeds can survive temperatures of up to 300°F (149°C). The plant has nitrogen-fixing bacteria in the root nodules, speeding its regrowth after a fire. Typical chaparral shrubs such as Manzanita and Chamise have waxy sheathes on their leaves to reduce transpiration and evaporative water loss in desert climate. When water stressed, their branches die, and strips of bark fall off. A mature plant can have up to half of its tissue dead, ready to burn. These plants catch fire readily and burn easily. The leaves contain highly flammable compounds such as terpenes, ethers, benzene, alcohols, fats, and resins, making the chaparral ecosystem a fire waiting to happen. The fires are not really destructive, since much of the plant (typically the underground parts), escapes the fire. Fire releases the nutrients that were bound up in the plant's dead tissues and recycles them. When the plants resprout from their roots, they grow much faster than they did before the fire. Without fire, the chaparral ecosystem would disappear, yielding its turf to the inexorable succession of scrub oaks.

Learning to Live with Fire

If fire is natural and beneficial in so many ecosystems, then what's the problem with it? As Walt Kelly wrote in Pogo: "We have met the enemy and he is us." Conflict arises in southern California (especially) and throughout the Mediterranean climate region (southern Spain, Portugal, Israel, etc.) between homeowners and the natural process of fire in the chaparral. We are living through a period of "desert chic." Chaparral is the "in place" to build a home, especially large and expensive ones, but the homeowners and developers are exposing themselves, their property, and their very lives to a high probability of fire. Unfortunately, humans and our homes are not as well adapted to fire as Manzanita and Snowbush. We can't bounce back from a fire as the plants do. So, there is tremendous pressure from homeowners in Southern California and elsewhere to suppress the very fires that created, foster, and maintain the unique natural beauty and biodiversity that draw them to the chaparral in the first place. If we suppress fire, the chaparral ecosystem will be lost forever.

Do people have the right to suppress fires that are essential to the survival of ecosystems? I don't think so. In the Bible (Genesis 1:26–28), we are commanded by God to have "dominion over all the fish in the sea, over the fowl of the air, and over every living thing that moveth upon the earth" (King James Version). Yet "dominion" does not suggest that we should be tyrants, ruthlessly exploiting all. Instead, we should be benevolent rulers, stewards of the Earth and all its inhabitants. We don't own the Earth, we share it with all other species.

When the homes in the chaparral burn, as they surely will, that is not a disaster but a tragedy. Human arrogance and ignorance of fire ecology of the chaparral brought on the destruction. "The fault lies not in our stars . . . but in ourselves" (Shakespeare, *Julius Caesar,* Act 1, Scene 2). We have created this conflict. Fire and the chaparral are eternally joined. To be good stewards, to preserve this exquisite piece of creation, we must allow it to burn. We simply cannot have chaparral without fire.

So, what can be done? The answer is simple: we must accept the vital role of fire as the natural disturbance it is and let the natural world live at its own pace, according to its own rules and requirements. We must stop developing the chaparral. If people insist on building there, the homeowners should absorb the risk and the costs. We could stop issuing fire insurance policies for houses built there, or they should carry high premiums to match the real risk. Since the chaparral has evolved to burn every fifteen years or so, the premium should equal one-fifteenth of the value of the house every year (that equals $150,000 annually on a million dollar home). Furthermore, communities could establish "no intervention zones" where firefighters will not risk their lives to try to save a home that should never have been built in the first place.

Preservation of a natural fire regime is essential to the preservation of all fire-dependent ecosystems, from the boreal forests, to the Lodgepole Pines and aspens of Yellowstone and the high-elevation forests of the American West, to the dwindling Longleaf Pine Savannas of the Southeast. We must find ways to live with fire. In chaparral, for example, the message is clear: "Keep Out!" In the remote mountains, with only few human inhabitants, there is generally little conflict. However, in the increasingly populated Southeast the conflicts are just heating up. What follows are some strategies for living with fire.

Prescribed Burns

The U.S. Forest Service and the U.S. Park Service have taken major steps in the right direction over the last few years. They now plan and implement controlled fires, or "prescribed burns." This prescription is real preventive medicine. By intentionally setting small ground fires, they seek to burn the undergrowth and reduce the accumulation of fuels on the forest floor, minimizing the potential for destructive crown fires. Prescribed burns should be timed to correspond to the reproductive requirements of the critical fire-dependent species. They are always planned around the weather, so the fire can, hopefully, be controlled. Issues of soil moisture, wind speed and direction, humidity, and approaching rain are all considered. The goal is to have a fire hot enough and big enough to have a salutary effect on the forest without burning out of control. It is a delicate balancing act, which regularly tests the knowledge and judgment of forest managers.

The new policy is called "Let it burn." Fire suppression is now being limited to fires that threaten life and property. Those that do not threaten human lives or property are allowed to burn themselves out. The "let it burn" policy, when applied to timber fires, is still controversial. In late spring 2000 the fire managers of Bandelier National Monument made a tragic mistake. The prescribed burn that they started blew out of control and became a wildfire. The result was that many homes were destroyed in neighboring Los Alamos, New Mexico. This conflagration raised cries of protest to ban prescribed burning. While the Los Alamos fire was a terrible error, foresters must not be deprived of this essential tool for ecosystem management; it is vital to preserving the health of forests worldwide.

After decades of social imprinting with the idea that fire is evil, it is hard to change people's minds. Fire is not evil; in fact, it is often good. Fire is essential for the survival and reproduction of many species. Without it, our world would be a much poorer, less diverse, less interesting place. It is time for Smokey Bear to retire. We need new images that reflect our new understanding of the value of fire. Perhaps a new symbol could be a peaceful grove of Quaking Aspens (call them "Quaker"?), gently sighing honor and respect for fire. Or perhaps a Red Cockaded Woodpecker ("Cocky"?) drumming out a celebratory song, praising fire for preserving its home.

Ecologically Based Zoning

Fires happen. Many ecosystems need fire the way a wheat field needs rain. With the "let it burn" policy becoming a national standard, the next thing we need to do is minimize the risk to people and property. To begin with, stop building in fire-hazard zones. We must recognize that it is in everyone's best interest not to build in fire-prone regions. We must pass ecologically based zoning regulations and ordinances. The goal is to protect our ecosystems from people (who cause most ignitions), and to protect people from "the fire next time." Since the fires will come as they must, the people need to stay out of their way. This will also save the social costs of the human devastation and rebuilding in the fire's wake.

Fireproof Building Codes

If people still insist on building in these ecosystems, then homes and businesses should meet tougher building code standards for fire resistance. Brick construction should be required and wood construction banned in these areas. Fire-resistant roofing materials (e.g., metal roofs instead of flammable asphalt shingles) and flame-retardant paint must be used. Metal studs, fireproof sheet rock, and concrete floors all contribute to minimizing the risk of fire. Each home should have a 300-foot buffer zone around it, free of all trees and fuels, as a permanent firebreak. Firebreak landscaping means that "the little cottage in the piney woods" will be relegated to nostalgic memory, but the reality of fire has made change necessary.

An ecological perspective that recognizes the inherent fire risk must be implemented in all planning and land management decisions. Insurance companies and actuaries need to come up with ecologically based fire insurance premiums. As the fire risk increases, the individual homeowner's premiums go up. When the fire risk is extreme, the cost should be prohibitive, encouraging people to build in places where the risk of fire is lower.

The message of the Lorax is to live in harmony with nature. I, too, speak for the trees, forests, and the ecosystems. We can all start by accepting fire as part of nature. This is an important step in building an ecological philosophy, an ecological worldview. Fire is a natural disturbance, and we must accept it, respect it, and even appreciate it. The payoff will be healthy, dynamic, and diverse ecosystems.

Further Reading

Chase, A. 1987. *Playing God in Yellowstone.* New York: Harvest Books, Harcourt Brace.

Fuller, M. 1991. *Forest Fires: An Introduction to Wildfire Behavior, Management, Firefighting and Prevention.* New York: John Wiley and Sons.

Taylor, M. A., and K. Steele. 2000. *Jumping Fire: A Smokejumper's Memoir of Fighting Wildfire in the West.* New York: Harcourt Brace.

Chapter 5 Disturbance Ecology and Flood Control

Old man river, old man river, he just keeps rollin' along.

—From Hammerstein and Kern, *Showboat*

I remember those lakes when they were rivers.

—Candy Feller, Smithsonian Institution, speaking of the conversion of Austin, Texas, rivers to reservoirs

June 1993, Southern Minnesota

The rains began mildly enough. It rained 12 inches the first day. Then, after a few days' respite, another 11 inches fell. As the rains continued, it became apparent that this was no ordinary storm. This was to become the wettest June in 115 years. As the waters of the upper Mississippi rose up out of its banks, people began to think of Noah. This was the beginning of the Great Midwestern Flood of 1993, the greatest, most extensive and most devastating flood in the recorded history of North America.

Ultimately, the floodwaters surged out of the banks along 500 miles from St. Paul, Minnesota, to St. Louis,

Missouri. The river was reclaiming its floodplain, having spread as far as 8.5 miles from its normal banks. Levees were breached. The Mississippi and the Missouri Rivers flowed together 20 miles upriver from their normal connection north of St. Louis. Decades of flood control and management were undone in a few weeks. The death and destruction were unprecedented. Twenty-four people died. Over 8 million acres were inundated by floodwaters. Property damage estimates ranged from $12 to $16 billion.

Floods are normal features of river and stream ecosystems. With a rainfall of that magnitude, there was no way to prevent this flood. Since a devastating flood in 1927 that had claimed 214 lives, the U.S. Army Corps of Engineers has tried to keep the Mississippi in its channel, building levees along hundreds of miles of riverbanks and dozens of dams. They built temporary holding ponds and diversion channels. They have invested over $25 billion on Mississippi flood control since 1930. Other state and federal agencies have spent billions more. This strategy has worked well enough for most flood events, but not for the big one of 1993. Did human attempts at flood control in the Mississippi contribute to the extent and intensity of the flood and the damage it caused? The answer is, unfortunately, "yes."

In an unconfined river, rising water spills out gradually into the floodplain. In the upper Mississippi, the levees confined all the extra water in the channel, greatly increasing its depth and velocity. So, when the water finally burst through, the river had much more destructive power. Instead of spreading the flow out gradually upstream, the levees held back the river until, much farther downstream, the flood burst through all at once. The power of the floodwater had far greater intensity and destructive force than if there had been no flood control at all.

The construction and success of the levees (funded to 80 percent of their cost by the federal government) changed people's attitudes toward the river: they thought it had been tamed and had lost their respect for the river's power. We built homes and towns in the floodplain. Beginning in 1968, people were further encouraged by federal flood insurance that would reimburse them for any flood losses. This insurance program was created to cover properties that private insurance companies considered too risky to insure. Nearly one half of the billions of dollars paid out after the 1993 flood went to people whose homes and property had been flooded before. These repeat flood victims were only 2 percent of the insured people in the area and they knew better than to live in the floodplain. Federal crop insurance had promoted living and farming too close to the river. So, our own policies made the damage worse. The underlying cause of this bizarre and self-destructive behavior is our lack of appreciation of flood dynamics and an inflated belief in our ability to control flooding.

Then, in the spring of 1997 (and again in April 2001), it happened all over again. The Red River rose from its banks and flooded Grand Forks, North Dakota. Waters filled the town. Boats were in the streets. This scene was repeated in Greenville, North Carolina, in the fall of 1999, when the rains accompanying first Hurricane Dennis, then Hurricane Floyd, flooded the Tar River. Photographs of the devastated small towns filled the news. Floods have seized the consciousness of the American public. We may have fooled *ourselves*, but "you can't fool Mother Nature": Old Man River just keeps rollin' along, and we need to learn to roll with it.

Why Do Rivers and Streams Flood?

Even small streams flood, often with dramatic effects. On February 17, 1998, I took my Field Ecology class to some local streams. The smallest stream, which is an unnamed tributary of Neville's Creek (my students named it "Reice's Creek"), was very shallow. Neville's Creek flows into Phil's Creek and then on into University Lake. My class had to climb down 1½ feet from the bank to reach the water. The streams were well inside their banks and easy to wade across. It rained steadily throughout our field trip, but later that night it poured. My rain gauge collected three inches of rain in seventeen hours. The next morning, all the streams were wild, and Reice's Creek was now a raging torrent. It had risen 3½ feet overnight and was way up, out of its bank. I walked down to Neville's Creek, where the floodplain is broad and flat. But the creek was now nearly 150 feet wide and flowing fast. I had never seen it so big or so violent. The transformation was astonishing. The previous afternoon, Neville's Creek had been flowing at about 8 inches (20 cm) per second. Now, it was flowing at about 9 feet (3m) per second, a 15-fold increase. It had 3-foot-high standing waves of muddy brown water, and it was roaring. These normally placid creeks had become raging, dangerous rivers.

The flood closed some low-lying roads in Chapel Hill, yet there was no damage to property. This was due to the wisdom of the Orange County (North Carolina) Watershed Protection Ordinance, which bans any land-disturbing activity, such as buildings, within the hundred-year floodplain of any permanent stream in the University Lake watershed. The county requires a minimum buffer zone of either 50 feet of trees or 100 feet of grass along all streams in the watershed. If any home had been built in that floodplain, it would have been washed away. Since there were no property losses, this flood didn't even make the news. The difference between our little local flood and the largest, most damaging floods like the Great Mississippi Flood of 1993 is only one of scale and degree of economic and personal loss. The processes involved are exactly the same.

Even with the most dramatic news coverage, we rarely get an explanation of why rivers flood. Let's start with some basic hydrology. Rivers and streams drain an area commonly called a watershed. Technically, it should be called a water catchment or drainage basin. Hydrologists use "watershed" to denote the ridge, which separates two drainage basins. However, the use of watershed is so entrenched in American usage that I will just "go with the flow." So, let's say that the watershed is the basin of land that captures rainfall. The water in that basin drains out through the stream, just like the water in a sink runs out through the drain. Analogously, we can think of a stream as the conduit for all the water in the drainage basin or watershed.

Rivers are organized as networks. Small streams start in low places that gather moisture or where subterranean springs emerge from the ground. This is called a *first-order stream.* One of the first principles of hydrology is—to no one's surprise—that water flows downhill. As first-order streams, typically little brooks, travel overland, they merge into larger and larger streams. Small streams flow together as we proceed down the watershed, with the streams increasing in order and size. Two first-order streams will merge and form a second-order stream. Two second-order streams will merge to form a third-order stream, and so on. Yet, stream order is an uncertain guide to stream size. While two first-order streams form a second-order stream, adding more first-order tributary streams doesn't change its order. Consequently, there could be a second-order stream with either two or two-hundred first-order tributaries, depending on the landscape. The huge and powerful Mississippi and Amazon Rivers are twelfth-order rivers. Naming streams is no clue to their actual size. Creeks and brooks are usually smaller than rivers, but there is no guarantee that this is always so. In North Carolina, New Hope Creek and the Rocky River are just about the same size. If you want to know how big a stream is, look at a detailed topographic map. The only reliable method of determining its size is to measure it.

Depending on the landscape, the watershed area (drainage basin) can be small and circumscribed or large and extensive. A stream's discharge is the total volume of water passing a given point on the stream bank per unit time, that is, the stream's width, multiplied by its depth, multiplied by the speed of the water, typically expressed as cubic feet per second (cfs). In recent accommodation to the metric system, the accepted unit now is cubic meters per second. The size of the drainage basin and the amount of precipitation in the watershed determine the discharge of a stream. All permanent streams have at least a nominal amount of water flowing, called the *base flow.* This is the stream discharge derived from groundwater, excluding all runoff from precipitation or snowmelt. *Bankfull discharge* is the amount of water that fills the stream without overflowing either bank. The stream can only hold so much water. When it is full, it spills out of its banks and a flood begins.

Both variation in rainfall patterns and watershed size will influence whether or not a flood occurs. A uniform rainfall of, say, two inches in twenty-four hours in a given region can cause flooding in a large watershed and none in a small one. If the basin is large, it is possible to have major rainfall in a portion of the headwaters, but very little downstream. The accumulated discharge of the tributary streams, which individually are at less than bank-full stage, can cause flooding downstream in the mainstem of the river. Another influential variable is the season. If a one-inch rain occurs during the growing season (spring to summer), then most of the water will be taken up by the vegetation and released to the atmosphere through evaporation and transpiration; very little surface runoff will occur and the river will not flood. However, the same amount of rain in the winter, when most plants are dormant and don't absorb the runoff, can result in flooding. Recall that, in 1995, just prior to Hurricane Fran, heavy rainfall in the previous week had saturated the ground's water-storage capacity. Consequently, all of Fran's seven inches of rain ran off the land into the streams and rivers, producing major flooding. The history of high rainfall in the Piedmont of North Carolina preceding Hurricane Fran made the resulting flooding much worse.

The discharge of a stream today, at any point, is the product of the accumulated rainfall throughout the drainage basin and subsequent discharge in the upstream tributaries over time. The result is that although the rain has stopped in an area, the river may continue to rise for several days. A flash flood occurs when the buildup of discharge travels downstream like a wall of water, which appears suddenly. Flash floods are extremely dangerous, since the local conditions may give no warning of the impending flood. When the water stops rising and begins to return to its banks, the flood has crested. The crest is the peak of the hydrograph (the plot of water height against time at a given point along the stream's length) and it moves downstream like a wave. When the crest passes a certain point like a town, it doesn't mean that the flood is over, just that the worst of it is over.

All floods are dangerous. Here's a simple warning. If you are driving and encounter a flooded roadway or bridge, turn back! Do not challenge the water. The power of floodwater can uproot trees, wash away permanent buildings, and tumble cars like dice. The strength of the swiftly flowing water can sweep your car into the flood. People die this way every year. Don't gamble that you can outrun a flood. The odds are against you.

Floods Are Natural Disturbances to Rivers and Streams

Floods are as vital to stream ecosystems as fires are to forests. Flooding disturbances enhance stream biodiversity. They prevent competitive exclusion, create the opportunities for recolonization, and allow streams to have the highest biodiversity of any freshwater ecosystems. Much of river management policy has been set by flood control and water resource agencies with the goal of preventing flood damage to cities, towns, and croplands. Nearly all rivers in the continental United States have been dammed. These dams also create reservoirs for water supply and recreation, but, although they were designed with the good of the human community in mind, they have created a spate of negative impacts on the natural aquatic communities of fishes and other animals and plants throughout the river's basin. As in forest management, we need to reexamine our flood control policies in light of contemporary disturbance theory.

For all their awesome and fearsome power, floods are normal and natural events in the history of rivers and streams. It is significant that the Fertile Crescent, the "Cradle of Western Civilization," included the floodplains of the Nile, Tigris, and Euphrates Rivers. Consider the Nile River. The winter rains in the mountains of Sudan feed the headwaters of the White Nile. In Ethiopia, the rains feed the Blue Nile. (The different soils and nutrients the rivers have scoured in the highlands cause the different colors of the White and Blue Nile.) These rivers join at Khartoum in Sudan to form the main channel of the Nile River. Ultimately, those mountain rains cause the springtime flooding of the river, which carries the nutrient-rich alluvium (the silt and clay) into the Nile Delta of Egypt. This flood has annually renewed the fertility of the soil, fostering the development of permanent settlements and agriculture and, ultimately, Western civilization.

How does flooding benefit people and ecosystems today? An aquatic ecologist, Peter Bayley, has recently summarized a new way of understanding the essential role of floods in ecosystem dynamics and health called the *flood-pulse concept*. The idea is richly ecological. It relates the dynamics of the river and the riparian (streamside) vegetation to the population dynamics and migration of fish and other aquatic organisms. The flood-pulse concept demonstrates that flooding increases biological productivity and maintains biodiversity in both the riparian zone and in the river itself. Bayley asserts that the regular flooding of riparian zones is not truly a disturbance, since the flood occurs predictably, usually in the springtime. I concur. In the debate over what constitutes a disturbance, several colleagues and I have argued that unpredictability is an essential element of disturbance. Many of the ecological benefits ascribed to flooding here are not strictly dependent on the regular timing of the flood.

Here is how the flood-pulse concept works. Let's take a look at it from the perspective of the floodplain, sitting high and dry, "waiting" for the flood (see fig. 5.1). The floodplain is the streamside zone that is alternately either immersed in river water or is dry land. The flooding may occur more than once a year. When the floodplain is in the dry phase, many nutrients are decomposed from the organic (bound) state into the mineral (free) state. As the river rises and inundates the floodplain, these nutrients are dissolved and released from the soil. Add all this to the dissolved nutrients in the river water, and you get a nutrient-rich soup. These free, inorganic nutrients are suddenly available to plants for growth, including emergent marsh and swamp plants (even trees) and submerged aquatic macrophytes ("big plants") and algae. The production of new plant tissue—primary production—is very high. Initially, decomposition cannot keep pace, resulting in a net increase in plant biomass. This pulse of new plant resources then fuels the rapid growth of the herbivorous invertebrates

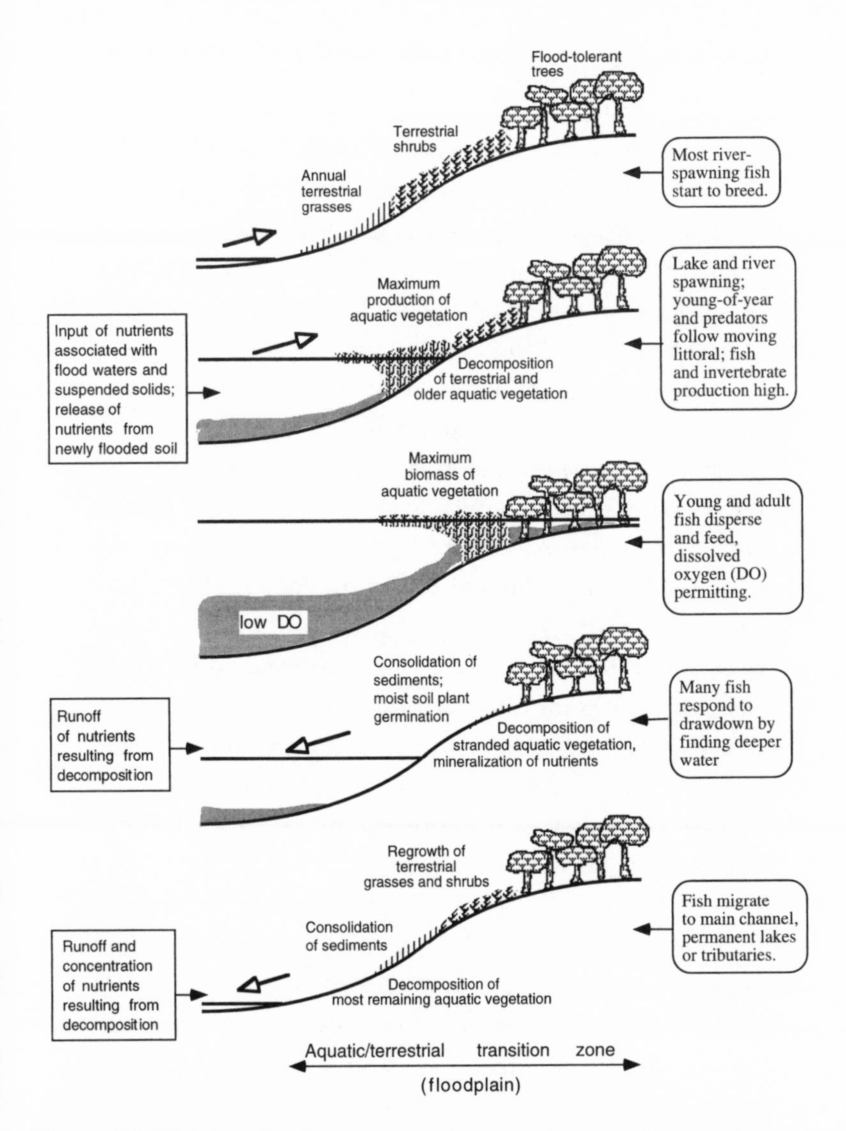

Figure 5.1 The flood-pulse concept: how regular flood cycles drive the processes in the riparian zone. (After Bayley 1995)

(mostly tiny crustaceans, or zooplankton, and larger aquatic insects). These, in turn, make a fine, hearty diet for spawning fish and the newly hatched baby fish. So, the period of rising floodwaters generates a pulse of increased productivity in the newly flooded zone.

As the rising river spreads into the riparian forest, these processes continue. As the water levels stabilize, the rate of decomposition (now aquatic) increases. The biological oxygen demand (BOD) increases because the massive number of bacteria and fungi (the lead group of decomposer organisms) must all respire to live. This respiration uses up much of the available dissolved oxygen. The increase in oxygen stress signals the fish that it is time to leave. As the river starts to draw down again, the fish go with it, following their need for dissolved oxygen. The river floodplain fauna—crustacea, insects, and fish—is well adapted to the short period of high resource abundance. They reproduce quickly, and young develop quickly. They exploit the resources and then get out.

As the river retreats, the newly dry, reexposed, nutrient-rich soil becomes a perfect place for the growth of terrestrial annual grasses. The result is an increase in the productivity of semiaquatic and terrestrial riparian zone plants in the wake of the dropping water levels. At the same time, the river water picks up an abundant supply of nutrients from the decomposition that took place on the flooded land. These nutrients drive the growth of suspended and attached algae in the water and submerged macrophytes on the river bottom. These plants become the base of the aquatic food chain.

The food chain starts with photosynthesis of the algae and macrophytes, which are eaten by herbivorous insects, which are in turn eaten by fish. Flooding therefore enhances both terrestrial and aquatic productivity. The cycle of flooding is essential to the life of every river.

How Human Land Use Intensifies Flood Damage

Wetlands are nature's buffers against flooding. Along the banks of rivers, streams, and lakes live wetland trees and other plants, which can tolerate and absorb large amounts of water during flood events. The water storage capacity of wetlands is dramatically reduced when the wetlands are channelized, drained, and filled for agriculture or development. In many areas of the midwestern United States it is common practice to install drainage tiles in fields to prevent water from standing in the fields and drowning the growing plants. Streams are straightened and channelized into drainage ditches, a practice designed to speed the drainage of water from the land. It works so well that a major cause of river flooding in the Midwest is tiling and ditching of agricultural fields. Before ditching, the water would be absorbed slowly by the floodplain soils and then flow gradually through groundwater pathways back to the river. But now, the water rushes off the land into the drainage ditches, pouring into the rivers at such a rapid rate that it can overwhelm the capacity of the river channel to contain it all. The result is flooding. Since these drainage practices have been installed, the rate of rise in the Mississippi River's water level has increased 22 percent, reflecting the surge of runoff into the main channel. The big rivers are much more likely to flood as a direct result of these agricultural drainage practices. Today, 95 percent of the Missouri River floodplain has been converted to row crop agriculture. Most of Illinois' streams have been converted to drainage canals. Such changes reflect a lack of appreciation of streams as living, vital ecosystems. Even the local farmers refer to these former streams as just "ditches."

Urbanization has the same kind of impacts. By increasing the proportion of the watershed that is paved over or covered with buildings and roads, we increase the impervious surface area. These surfaces cannot absorb the rainfall, so the water rushes off the land, into the storm drains, and on into the rivers. Streets and roads can become conduits for runoff water. Consequently, urbanized streams flood more frequently and violently than rural streams in similar-sized watersheds.

Stanley Riggs of East Carolina University recently made an important observation, showing that urbanization and road construction change the drainage patterns of runoff water. Riggs pointed out that roads, bridges, and overpasses create partial barriers to the flow of the stream or river they cross. When we install culverts to carry a stream under a roadway, they are often too small to carry peak flows from major storms or hurricanes. When a stream or ditch, carrying runoff from the land, encounters these barriers, the water has no place to go except to spread out and flood the surrounding area. Riggs commented that during Hurricane Floyd in 1999, "Ahoskie [N.C.], which is high above any river, flooded because ditching and draining moved more water off the land faster, which was then held back by road dams." The worst flooding occurred in Princeville, North Carolina, because the new Highway 64 bridge formed a road dam that forced the Tar River out of its banks. In the absence of these "road dams," the water would have drained naturally with minimal damage. These events are tragedies, in the classical sense, because they could have been avoided and we caused them.

You Can't Fool (with) Mother Nature

Egypt is a desert country. The Nile is Egypt's only major source of freshwater. Throughout history, Egypt's farmland was limited to the Nile River's floodplain, a narrow strip 3–6.25 miles (5–10 km) on either side of the Nile. The historical main agricultural area has always been in the Nile Delta. The normal, undammed flood pattern of the Nile was always the typical Mediterranean climate pattern: winter rains, spring floods, and summer dry conditions. The floodwaters replenished the soil and nutrients in the Nile Delta, making it a fertile and productive agricultural area.

In an attempt to tame the Nile and create a reservoir to irrigate more farmland, the former USSR built the

Aswan High Dam in the 1970s, one of the world's largest dams. The Aswan High Dam is 375 feet (114 m) high and 3,600 feet (1,811 m) long. It was built for water storage, irrigation, and hydroelectric power generation. The water stored behind the dam in Lake Nassar is sufficient to irrigate more than 7 million acres. With the giant impoundment of Lake Nassar, irrigation was possible all year, greatly extending the growing season. Water releases from the dam kept the Nile flowing year-round. It also effectively ended the cycle of spring floods and summer drought of the Nile River. It is an archetypal example of flood control gone wrong.

The consequences were completely unexpected. Soon after the Aswan High Dam was operational, a massive outbreak of schistosomiasis (also known as Bilharziasis or snail fever) took place in the region around Lake Nassar. Schistosomiasis is a human disease of the liver caused by parasitic blood flukes of the genus *Schistosoma.* The intermediate host of the *Schistosoma* larvae is a snail, which lives in the Nile. As the larvae mature, the snails release them into the water. There they can penetrate the skin of people who come in contact with the infested water. People commonly bathed in the new lake and did their laundry there and became easy victims of schistosomiasis. An epidemic ensued. The mature worms pass out of the human hosts in fecal matter. When the feces reach the water, the worms infect the snails, thus closing the loop. So, poor sanitation drove the epidemic.

What is the relationship between the schistosomiasis epidemic and the Aswan High Dam? Before the dam, the natural fluctuations of the Nile's water level caused the Nile floodplain to dry out in the summer and fall, causing major snail mortality. This naturally lowered the density of the intermediate host for schistosomiasis. After the dam was built, the lake inundated the floodplain, keeping it wet all year, so the snail population increased to record high levels. Rather than hitting a seasonal bottleneck as their habitat shrunk each summer, the snails were released from the natural hydrological and ecological constraints and the snail population exploded. Since the water was available year-round, people used it year-round, too, further enhancing their exposure to *Schistosoma* and increasing the probability of infection. By changing the Nile River's normal annual cycle of flood and drought, the Aswan High Dam led to an ecological release for the snail population, creating the schistosomiasis epidemic. In 1935, only 5 percent of Egyptians had schistosomiasis. By 1972, the number had exploded to 35 percent. Outbreaks of schistosomiasis have followed dam construction throughout subtropical Africa. The World Health Organization developed a snail control program, which finally checked the epidemic in Egypt.

The ecological problems of the region continue today. The nourishing silt transported by the Nile, which replenished the Nile Delta for millennia, is being trapped behind the Aswan Dam in Lake Nassar. Without flooding, the delta is losing its natural fertility and is even shrinking. The disturbance in this case was human-induced elimination of the natural flood-drought cycle.

Real-World Examples of the Benefits of Flooding to Stream Ecosystems

The Great Mississippi Flood of 1993

Flooding is an essential recharge mechanism for rivers and their floodplains. The timing of floods and their variations are critical to the maintenance of biodiversity. The beneficial effects of flooding are often manifest in the response of fish populations to spring flooding in rivers. In the Mississippi River basin, the loss of streamside forests to agriculture, the draining of wetlands (vital spawning grounds for fish), and the extensive system of levees and dams has done major damage to the fish community. The commercial fish catch has fallen by 83 percent over the last fifty years. Paddlefish (*Polyodon spathula*), one of the most evolutionarily ancient, primitive fish native to the Mississippi, were a valuable commercial catch in 1960, weighing up to 160 pounds (72.6 kg) each. By 1980, their very existence was threatened. At one spot, the 1960 catch of Paddlefish was 16,000; by 1980 this number had dropped to 125. Human intervention had deprived this and other native species of their spawning grounds. Levees and dams blocked their migratory routes. When these were overtopped or breached by the floodwaters, the breeding fish could, at last, reach their ancestral spawning grounds along the banks. In the Mississippi River, stimulated by the Great Flood, 1993 was a record spawning year for native fish.

Dams and Controlled Flooding of the Colorado River in Grand Canyon National Park

At 1,450 miles (2,334 km), the Colorado River is the longest river west of the Rocky Mountains. The Colorado carved the magnificent Grand Canyon in northern Arizona, one of our national treasures. The canyon is 1 mile deep, 9 miles wide, and 217 miles (349 km) long. It was chiseled into the sandstone and other rocks of the Kaibab Plateau by the slow erosion and downcutting of the river by the normal flood events of the Colorado River. These floods were driven by melting snows in the springtime and by massive thunderstorms in the summer. The thunderstorms still boom across Arizona today, but their power to recarve the Canyon has been negated by flood control dams. Spring floods are just a memory in the Grand Canyon. To control its tremendous destructive power, dams have been built along the Colorado River and its tributary streams. The U.S. Bureau of Reclamation built many—with the noblest of goals—and it succeeded in totally squelching the flooding of the mighty Colorado. Notable is the giant Hoover Dam, 726 feet (221 m) high and 1,244 feet (379 m) long. The Hoover Dam impounds the river downstream of the Grand Canyon to form Lake Mead, one of the largest artificial lakes in the world. Upstream of the Grand Canyon is the Glen Canyon Dam. It forms Lake Powell, named for American geologist John Wesley Powell, who explored the Colorado and was the first, in 1869, to take a boat through the Grand Canyon. This system of Colorado River reservoirs has a storage capacity of 2.6 times the volume of the annual discharge of the river. The benefits of these dams, besides regulating the floodwaters, include hydroelectric power generation and water storage reservoirs for irrigation projects and recreation. They have made agriculture possible in this arid region.

So, what's the problem? Hydrologist John Schmidt and his colleagues have analyzed the problems of the Colorado River. It took approximately ten million years for the Colorado to cut the Grand Canyon. Now, due to damming of the river, it hardly floods at all, and major floods are a thing of the past. The dams have reduced the frequency, size, and duration of floods. The biggest recorded flood on the Colorado was measured by the U.S. Geological Survey at 3,550 cubic meters per second (127,000 cfs) on July 2, 1927. Estimates based on high-water marks suggest that flows of up to 8,400 m^3/s were not uncommon. Before the construction of the Glen Canyon Dam, the peak two-year flood (the greatest flood in a two-year span) was 2,150 m^3/s, and flows above 1,400 m^3/s lasted for more than thirty days a year. Since the dam's construction, the two-year peak flood was only 679 m^3/s, and flows greater than 1400 m^3/s occur just three days a year. The Colorado River has been deprived of its natural flood cycle.

The Glen Canyon Dam also cut off essential sources of sediments. Before dam construction, at Lee's Ferry, upstream from the Grand Canyon, sediment loads were 13.6 million tons/year (6×10^{10} kg/yr), and the floods built debris fans and sandbars in the Colorado. The rapids in the canyon were flushed of sediments, leaving only the biggest boulders. Since the dam was built, virtually no suspended sediment has been found in the Colorado at Lee's Ferry. Sediments still enter the river from the Little Colorado and Pariah Rivers, but only $4.1. \times 10^5$ tons/year (1.8×10^{10} kg/yr). Now, due to the slow flow, the modest amount of sediment suspended in the river settles out and accumulates in the canyon's river bottom, clogging the spaces between the rocks.

The third problem caused by the Glen Canyon Dam is the change in the thermal regime of the river. Like all normal rivers, the Colorado is cold in the winter and warm in the summer. However, since nearly all the water of the Colorado is stored in Lake Powell, the lake controls the temperature of the river water. Lake Powell is very deep, so the water at the surface gets warm but floats on a layer of colder (and heavier) water. This is called *thermal stratification.* Even in midsummer, the deep water is cold. The release point for the river water is in this deep, cold part of the lake. So, the depth of the discharge point in the dam determines the temperature of the river water. Presently, the water released is a nearly constant 46.6–50°F (8–10°C). This cold water is also nutrient rich and higher in salt content than the warmer surface waters of Lake Powell, which get up to 86°F (30°C).

What does all this mean for the native fish of the Colorado River? Before we started introducing various game fish around 1880, 74 percent of the fish species were endemic (endemic species are native species found exclusively in a given area). The Colorado River had the highest level of endemism of any North American river basin. These species evolved in the ancestral hydrologic regime of the Colorado River, which included spring flooding, high sediment loads, high turbidity, and warm summer temperatures. The lake and dam have altered all of these essential environmental conditions. The altered flow and thermal regimes have translated into a crisis for the native fish. Eight of the thirty-five native fish species of the Colorado River that were found in the Grand Canyon before the construction of the Glen Canyon Dam are now either extinct or endangered. The Colorado Squawfish (*Ptychocheilus lucius*), the Bonytail (*Gila elegans*), and the Roundtail Chub (*G. robusta*) were all extinct in the canyon by the 1970s. Five native species still exist in the upper basin, but the Razorback Suckers are endangered. The constantly cold water, which now flows in the river, is too cold for the native species to reproduce. The lowest temperature that permits these species to spawn successfully is 61°F (16°C), but now the river hardly ever gets that warm. Three of the native species—the Speckled Dace (*Rhinichthys osculus*), the Bluehead Sucker (*Catostomus dicobulus*), and the Flannelmouth Sucker (*C. latipinnis*)—do reproduce in the tributaries, but reproduction of the native species in the mainstem of the Colorado has ceased. The sediment deposition compounds the fish's problems. Sediments clog the spaces between the rocks and smother the gravel beds that are necessary for the breeding and nesting of the native fish species. Consequently, reproduction of the native Colorado River fishes in the Grand Canyon has stopped.

A further threat to the native fish comes from the introduction of exotic species (nonnatives), including Carp (*Cyprinus carpio*), Channel Catfish (*Ictalurus punctatus*), and Fathead Minnows (*Pimephales promelus*). An even greater threat comes from introduced cold-water trout species—Rainbow Trout (*Onchorhynchus mykiss*), Brown Trout (*Salmo trutta*), Brook Trout (*Salvelinus fontinalis*), and Cutthroat Trout (*O. clarki*). These species are well adapted to the cold water conditions and reproduce freely. These trout are strong competitors with the native species and prey upon them. The endemic fish species of the Grand Canyon reach of the Colorado River have their backs (backfins) against the (Canyon) wall.

The deposition of sediments and high nutrient conditions in the canyon bottom have created the opportunity for the development of extensive marshlands on banks that had been kept free of vegetation before the dam by the scouring action of the floods. The ancestral high waters had also been a seed source for the upper riparian zone, which has now been cut off from the river. The upper riparian zone vegetation is in great decline because of drought stress. In an ironic twist to this story, the lower riparian marshlands have been colonized by two endangered species, the Southwestern Willow Flycatcher (*Epidonax traillii*) and the Kanab Ambersnail (*Oxyloma haydeni kanabensis*). The entire

Colorado River ecosystem has been changed and damaged by the operation of this system of dams. All of these problems, including the threats to the native fish, derive from the absence of flooding and the elimination of the seasonal fluctuations of water level, flow rate, and temperature. The damage is being done by the elimination of disturbance, paralleling the damage to forests caused by fire suppression.

In late March and early April 1996, an attempt at remediation took place in the form of a release of water from Glen Canyon Dam at a rate of 1,400 cubic meters per second to create a controlled spring flood. The management goal of this flood was to rebuild beaches and restore fish habitat in the Grand Canyon. While this seems like a bold step at river restoration, in fact it was quite modest: it was the post-dam equivalent of a ten-year flood, and it caused less discharge than the pre-dam peak two-year flood. Several limitations to this controlled flood are worth noting. First, the flood was of cold water (yet not as cold as a normal spring flood, which would be driven by snowmelt). Second, it came earlier in the springtime than a natural spring flood. Third, since the water came from Lake Powell, it was essentially sediment free. As such, it could not adequately renourish the sediment-starved beaches. Fourth, it was a short-duration event, lasting less than ten days. The 1957 pre-dam spring flood began in May, peaked at 3,500 m^3/s and exceeded 1,400 m^3/s for about two months. Still, the controlled flood did some good. It did scour some sand from the channel bottom and redistributed it to the stream bank, thus rebuilding some beaches (re-creating some campsites, which had been lost since the dam was built). However, for the native fish it is a classic case of "too little, too late." The fate of the native fishes of the Grand Canyon reach of the Colorado River is a conundrum. There is no easy right answer.

There are multiple trade-offs between costs and benefits. Should we offer protection to the Razorback Sucker or the Kanab Ambersnail? Should we tear down the Glen Canyon Dam and restore the river? If we do, then we lose the hydroelectric power and the irrigation water. Is there a compromise, one that protects the native fish and allows irrigation and power generation to continue? One compromise, suggested by the National Research Council, is the "simulated natural ecosystem" approach, which puts unvegetated sandbars and endangered fish on top of the priority list. It would require dam operators to release water to mimic the pre-dam hydrograph instead of following the high power demands that drive midsummer releases now. It could require reengineering of the water release point to bring it higher in the dam, to raise the temperature of the water released in summer. There would also need to be a way to pump sediments up into the lake's discharge; this would not only help the fish but would also have the additional benefit of lengthening the useful life of Lake Powell, which is presently being filled in by Colorado River sediments. The key point is that our attempt to control Colorado River flooding with the Glen Canyon Dam has proved incompatible with the survival of the natural fish community of the river. The natural disturbance regime of flooding and subsequent low water are essential for the continued survival of the Colorado's remaining native fish.

DAMMING OR DAMNING THE WEST: DESTRUCTION OF THE COLUMBIA RIVER SALMON RUNS

The conflict between dams and river ecosystems is ubiquitous. A classic example is the destruction of the Pacific Salmon fishery. The culprits of this disaster are now becoming clear. A major factor is the damming of the Columbia River system. (The decline of the Columbia River salmon run is a story of extreme complexity. Logging, development, and overfishing all have played a significant role. Here I will focus on only one aspect, the effects of dams. The Columbia River is about 1,245 miles (2,005 km) long. After receiving the Snake River, the Columbia turns west and forms much of the boundary between the states of Washington and Oregon before emptying into the Pacific Ocean through a broad estuary. The river has carved magnificent canyons and valleys. About one-third of its length is in Canada. The Columbia and its tributaries drain a huge watershed—260,000 square miles (673, 400 km^2).

Pacific Salmon is a complex of six species of the Genus *Onchorhynchus*. These include the Chinook (*O. tschawytscha*), Chum (*O. keta*), Coho Salmon (*O. kisutch*), Pink Salmon (*O. gorbuscha*), Sockeye or Red Salmon

(*O. nerka*), and Steelhead (Rainbow Trout, *O. mykiss*). The Columbia River salmon run drives an industry valued at billions of dollars. The salmon catch peaked in 1883, when lock and dam construction began in earnest. Salmon are anadromous fishes, meaning they use both freshwater and marine ecosystems to complete their life cycle. The adults spend from several months to several years swimming, feeding, and growing in the Pacific Ocean, with the duration of the oceanic phase varying with species. Then, in one of the wonders of nature, the adult salmon return to the stream of their birth. It is incredible that, after as many as four years of feeding in the Pacific Ocean, they can find their natal stream. They migrate back up the river to mate and lay their eggs. Then the adults die. Their subsequent decomposition is a vital source of nutrients for the whole river ecosystem. Bears and raccoons eat the salmon carcasses, thus transferring the salmon's nutrients (especially nitrogen) to the forest when they excrete. In this way, nutrients gathered in the ocean are transported all the way up to the headwater forests. The eggs hatch, and then the fingerlings grow and feed. They swim downstream through larger and larger streams and rivers and ultimately out to sea. The duration of their freshwater residence ranges from a few days for Chum Salmon to as much as four years for Coho Salmon.

The massive migration of the salmon back up the rivers has supported a fishery for centuries. First, it consisted only of brown bears standing in and beside the streams fishing for salmon. Then bears shared the catch with Native Americans. Now, a recreational and commercial fishing industry has become a vital economic force in the Pacific Northwest. But, the salmon are disappearing. The damming of the Columbia and Snake Rivers (the latter is the Columbia's main tributary) began in the 1930's. The complex of twenty eight major dams along the Columbia River drainage blocks the salmons' passage to their breeding grounds. The best known of these is the Grand Coulee Dam in Washington. Its height of 550 feet (167.5m) poses a massive obstacle to the migrating salmon. Since the most recent dam construction in Canada in the 1980s, the Columbia River hasn't flooded at all. The goals of the dams in the Columbia River system include river navigation (through a series of locks and dams), hydroelectric power generation, flood control, and irrigation for agriculture. The diffuse effects of the dams are much the same as in the Grand Canyon. The dams have altered the cycle of flooding and robbed the rivers of their sediments. Did the streams' sediment plumes stretching out into the ocean provide the vital clues telling the salmon where their natal stream is? If the sediments are gone, how will the fish find their homes? Although millions of dollars have been spent on fish ladders and bypass canals and other methods designed to give the salmon a way past the dams, the spawning runs are but a minuscule fraction of the millions and millions of fish witnessed by Lewis and Clark in 1805. It would be naive, and unfair, to blame the demise of the Pacific Salmon on the dams alone. Other forces have been at work.

Clear-cutting of the forests has led to massive erosion and sedimentation of the streams of the Pacific Northwest. The recent construction boom in that region has had a hand, too. Land clearing for subdivisions and shopping centers leads to sedimentation of the streams. The sediments fill in and bury the gravel beds that the salmon need to build their nests and protect their eggs. Sedimentation can smother the young fish fry and fingerlings, and sedimentation problems are made worse by the dams. Natural floods, typically driven by spring snowmelt from the mountains or summer thunderstorms, scoured out the fine sediments (the silts and clays). Floods wash gravel beds clean of silts, making them hospitable as salmon nurseries. With the dams preventing flooding, the sedimentation problem just builds up, further endangering the survival of the salmon.

Flooding Is Essential for Healthy Rivers and Streams

Rivers are flowing waters, and they will flood. Dams are designed to prevent flooding, to "protect" us from the natural flood disturbance of rivers. They block the flow of rivers and fill reservoirs. Dams are fundamentally antagonistic to the natural functioning of rivers. The whole aim of building a dam is to create a lake where a river used to run free. They trap sediments and nutrients. They distort (and usually flatten) the natural hydrograph. They prevent most floods, but make others more severe.

Damming has been successful from an anthropogenic perspective. We have tamed most floods and harnessed the power of the rivers to generate hydropower. We have diverted the flow of mighty rivers into manmade lakes for irrigation and recreation. Cities need the hydroelectric power and the water supply, and reservoirs are important recreational resources. We have made the rivers serve us. The benefits to people are clear, but what about the costs? What is a healthy salmon fishery worth? More important, what is a healthy Columbia River or Colorado River ecosystem worth? We have eagerly reaped the benefits, but now we are beginning to see the negative consequences. Our rivers and streams are showing the strain, and we are losing our native species. Nature has had enough.

It is the nature of rivers to flood. This natural disturbance, the flood, is essential for the health of rivers, streams, riparian forests, floodplains, wetlands, and fisheries. In our rush to control nature, we are harming our ecosystems and ourselves. *The nature of nature is change.* Can we simultaneously dam our rivers and protect them? The demise of the Pacific Salmon is pitting the interests of fishermen and the fishing industry against agriculture, forestry, and development. The lock and dam system of the Mississippi is trading a healthy ecosystem and fishery for agriculture and barge traffic. The flood control on the Colorado provides irrigation but is dooming an entire ecosystem. We cannot continue on our present course. Streams and their floodplains and wetlands give us more than fish. They provide clean water. They decompose and neutralize our wastes. They rejuvenate our lands. They provide us with beauty and tranquillity. All of these services are provided to us for free. To preserve these vital functions of rivers we must reinstate the natural disturbance regimes. Natural floods are integral and vital to the very existence of stream ecosystems. We must learn to live with floods to protect and preserve our stream ecosystems and to preserve their biodiversity.

Further Reading

Ball, P. 2000. *Life's Matrix: A Biography of Water.* New York: Farrar Straus and Giroux.

Bayley, P. 1995. The flood-pulse concept. *BioScience* 45: 153.

Cone, J. 1995. *A Common Fate: Endangered Salmon and the People of the Pacific Northwest.* New York: Henry Holt.

Dean, C. 1999. *Against the Tide.* New York: Columbia University Press.

Leopold, L. B. 1997. *Water, Rivers and Creeks.* Herndon, VA: University Science Books.

Resh, V. H., A. V. Brown, A. P. Covich, M. E. Gurtz, H. W. Li, G. W. Minshall, S. R. Reice, A. L. Sheldon, J. B. Wallace, and R. C. Wissmar. 1988. The role of disturbance in stream ecology. *Journal of the North American Benthological Society* 7: 433–455.

Schmidt, J. C., R. H. Webb, R. A. Valdez, R. Marzolf, and L. E Stevens. 1998. Science and values in river restoration in the Grand Canyon. *BioScience* 48: 735–748.

Sparks, R. C., J. C. Nelson, and Y. Yin. 1998. Naturalization of the flood regime in regulated rivers. *BioScience* 48: 706–720.

Chapter 6 ***Biodiversity, Ecosystem Services, and Human Needs***

Variety is the spice of life!
—Anonymous

Yellowstone National Park, June 15, 1998

Ten years had passed since the fires of 1988. The burned hulks of Lodgepole Pines shrouded the hillsides. However, if you looked closely now, you would see a green carpet beneath the dead and charred trees. Below these skeletons, thousands upon thousands of 18-inch-high Lodgepole seedlings are sprouting. The forest is coming back.

What is the value of Yellowstone National Park? How do we measure the worth of any ecosystem? We could measure Yellowstone's worth by calculating the sum of all the admissions fees, plus the dollars spent at lodges and restaurants and concessions. We can and do measure the revenue from tourism in the area. Yet, that calculation falls short of the true value of Yellowstone. How do you place a dollar value on one of the last free, wild buffalo herds in the world? What is the value of seeing a geyser erupt, watching a cascading waterfall, or hearing the song of a Western Meadowlark or the howl of a Gray Wolf? We pay scant attention to commodities that cannot be priced, such as clean air or fresh, clean water. Over one hundred years ago, Henry David Thoreau had proclaimed: "In wildness is the preservation of the world." He declared this out of his faith in nature and his understanding of the relationship between wildness and humankind. Now, through our studies of ecosystems, we can begin to place value on what wildness does for the world. We are learning just how right he was.

What Is Biodiversity?

Biodiversity is one of the most important ecological concepts to have entered the public consciousness in the last decade. Everyone is for preserving biodiversity. Images of biodiversity, from whales and tree frogs to endangered pandas, adorn everything from T-shirts to milk cartons. Biodiversity has attained an almost sanctified status, akin to motherhood. Most public attention to biodiversity is devoted to species diversity. The Endangered Species Act has established the preservation of species and, as a result, biodiversity, a federal law.

In the plainest sense, biodiversity is the variety of living things. Recall that species richness is the total number of species in an area. We can refer to the species richness of New Hope Creek, North Carolina. If we totaled up our species list, we would discover that we have 256 distinct species of macroninvertebrates there. We can compare that to another stream, one that is heavily polluted and contains only ten species. Species richness is the most straightforward assessment of biodiversity, and the most widely used. Environmental agencies (such as the U.S. Environmental Protection Agency), interpret species richness as a sign of ecosystem health. Remember that there is a vital link between disturbances and biodiversity. Joseph Connell's Intermediate Disturbance Hypothesis has shown that a moderate level of disturbance leads to the highest species richness. So, moderately disturbed ecosystems have high biodiversity, which is an indicator of ecosystem health.

Species richness is often thought of as synonymous with biodiversity. However, in fact, it is only one part of the whole biodiversity problem. At a higher level, we have the diversity of community types across the landscape. At lower levels, we have diversity of age structure of single-species populations and genetic diversity within populations. All of these are important components of the overall biodiversity of the planet. Ecologists concerned with community structure go beyond species richness and consider how evenly distributed the individuals are among the species in the community. If 98 percent of all the insects in New Hope Creek consisted of just one mayfly, then the stream would contain a nearly homogeneous monoculture. If, instead, all 256 species were equally abundant, the stream community would be far more diverse. To fully appreciate the diversity of a community we must weigh both species richness and the evenness of distribution of individuals among species.

Imagine that you are walking in a tree farm in the North Carolina mountains, where they are growing Fraser Firs (*Abies fraseri*) for Christmas trees. Every tree you encounter will be a Fraser Fir. You could close your eyes, spin around, and the next tree you touch will again be a Fraser Fir. If, in contrast, you are walking in the Amazonian rainforest of Brazil, which has one of the most diverse floras on Earth, all the trees will seem to be different from one another. There are over two hundred species of trees. Looking carefully, you may walk a long distance before you see the same tree species a second time. The rain forest has both high richness and high evenness, and consequently high biodiversity. Therefore, you have only a small chance of knowing which species you will encounter next. There are many ways to calculate species diversity, and all diversity measures are not alike. An index that incorporates both richness and evenness tells you far more about the community structure than species richness alone. The most popular one is the Shannon Index of Species Diversity.

Disturbances act to remove individuals and permit recolonization. For example, treefalls and storm damage cause disturbances in rain forests that open the forest to recolonization by a great variety of species. Disturbances tend to reduce the abundance of the most common (dominant) species and thus increase the evenness part of biodiversity. The top competitors are replaced by some colonizing species, increasing the richness aspect of biodiversity. Disturbances and recolonization enhance all aspects of biodiversity in ecosystems.

Why Is Species Diversity Important?

Biodiversity works in many ways to stabilize populations and ecosystem-level processes. Why, for example, do predators coexist with their prey? Why don't they eat them all up, driving the prey to extinction? If a predator has only a single prey species (i.e., if Red-tailed Hawks ate only one species of field mouse) and a disease wiped out the mouse population, then the predator would starve and go extinct. If, however, the predator has many species of prey to choose from, it would be far less vulnerable to the fluctuations in any single prey species. In fact, Red-tailed Hawks eat a wide variety of prey species. They will eat almost anything they can see and catch. They eat field mice, voles, squirrels, and rabbits. If a virus wiped out the mice, the hawks would still thrive. Most predators have very broad diets. Variety is not only the "spice of life" but helps preserve life itself.

The landscape is a mosaic of different ecosystems. Each ecosystem type has a unique array of species in its community, and each contributes something special to the local and global environments. Wetlands, which we derisively used to call swamps, are valuable for their ability to absorb and sequester excess nutrients, thus providing a protective buffer to the adjacent streams and lakes. Rain forests, formerly called jungles, are now fashionable for their high species diversity and their critical role in the global environment. Rain forests help maintain the global supply of oxygen through high rates of photosynthesis. The variety of ecosystems and their communities and individual species are of value to humankind for the ecosystem services they provide. Each has a role in the complex living biosphere, and none is expendable. The interdependencies of one ecosystem on another, one species on another, and each individual on another in the web of life is reason enough to treasure biodiversity. Here's a reminder: natural disturbances guarantee continued high biodiversity. That is their value, their "silver lining."

How Ecosystems Work and Why Biodiversity Matters

Why is species diversity so often used as a measure of ecosystem health? All species are important to the functioning of the ecosystem in which they live. We cannot know which ones are expendable until they are gone. We can try to pick and choose among the species, but we would do that at great risk, since we don't fully understand how each species fits into the web of life. I like bluebirds better than slime molds, but which is really more important? Recently, ecologists have begun documenting the value of species diversity to the ecosystem. All ecosystems have a set of functions. In this section, I will describe what ecosystems do and the vital role of biodiversity in performing those functions. Then we will examine specific ecosystem services (such as cleaning air and water) that are essential for all humankinds.

The functions of all ecosystems fall into three broad categories of processes that occur in all ecosystems. All ecosystems must (1) acquire energy, (2) process and use that energy, and (3) recycle nutrients. All life requires a source of energy. Most of that energy enters the ecosystem through photosynthesis by green plants. In photosynthesis, the energy of light from the sun is captured by chlorophyll (the molecules that make green plants green). This energy is used to combine carbon dioxide (CO_2) and water (H_2O) into energy-rich carbohydrate molecules such as glucose ($C_6H_{12}O_6$). The carbohydrates are energy packages, which fuel all living processes. Photosynthesis is the key to life. This process is called *primary production* since it is the first phase of production of living tissue. Even dark systems like cave ecosystems or the deep ocean floor get their energy from organisms that photosynthesize in the light. Leaves (and animals) are carried into caves by streams, and dead algal cells (and plankton) drift down from the ocean surface to the depths. In every ecosystem, all other processes depend on first acquiring energy.

Second, all ecosystems must have a way to use the energy captured by plants. Plants alone do not make an ecosystem. There must be a way to use the energy and materials they produce. That's where animals come in. Animals eat the plants and convert the plant matter and energy into animal tissues. Animals eating plants (and animals eating other animals) makes the energy and materials in the prey organisms available to the predators. Decomposing organisms, notably the bacteria and fungi, can break down the dead plants and animals, and use them for growth. That way, the energy is not trapped in dead wood and other plant tissues but can be utilized by the rest of the community. A food chain is the sequence of who eats whom. The links in a food chain are the species that eat one another in turn. All foods are derived from plants' photosynthesis. Organisms convert their foods (whether from animal or plant tissues) into a usable form of energy (adenosine triphosphate molecules, or ATP) through respiration. ATP powers every known function of animals and plants from growth to motion to thought.

The third major function of ecosystems is nutrient cycling. Nutrients are the building blocks of all living things. These include the nitrogen, phosphorus, carbon, and other substances that all cells are made of. Every ecosystem must recycle nutrients. Imagine a forest where nothing lived but trees. The trees would draw the nutrients out of the soil and grow. Since no organism can live forever, the trees would eventually die. All the nutrients would be bound up in the dead tissues and nothing could ever grow again. Without nutrient cycling, life as we know it would cease. Nutrient cycling is present in every ecosystem—it is essential for the continuity of life. Decomposition, a complex process whereby the dead tissue is broken down into its constituent parts, is a key process in nutrient recycling. Decomposer bacteria and fungi feed and grow on dead organic matter. Animals then eat the bacteria and fungi. Without decomposition, dead leaves and branches would just pile up on the forest floor, ultimately burying every living thing. Through the action of bacteria, fungi, and animals, the water and nutrients bound up in the organic matter are returned back to the soil. CO_2 and water return to the atmosphere. In their now free state, these molecules are available for new plant growth. Decomposition is the quintessential recycling mechanism. It may look or smell nasty, but it is an essential part of the natural functioning of all ecosystems—terrestrial, freshwater, and marine—and is the link between death and new life. It is one of the essential services that ecosystems provide free of charge. All ecosystems do all of these three jobs.

Ecosystem Services, or What We Get from Ecosystems for Free

Food

Primary production is the ultimate source of all of our food, from rice and beans to beef and shellfish. We eat plants directly (lettuce and spinach). We eat plant products such as fruits (apples and oranges) and their seeds (wheat and rice and beans). When we eat beef, we are eating cattle, which eat grass and grain, so we are eating plants indirectly. When we eat seafood, the food chain begins with green plants (typically microscopic algae) that are eaten by tiny animals (called zooplankton). These, in turn, are eaten by very small fish (minnows), which are eaten by bigger fish (sunfish in lakes, spot and croakers in the Atlantic Ocean). They in turn are eaten by yet bigger fish (bass or tuna), which people eat. The entire food chain is dependent on the tiny algae, the primary producers, which captured the sun's energy and made food available to us. This ecosystem service is of fundamental importance to us. Without primary production we would simply starve. In our own self-interest, then, we must protect the integrity of the ecosystems—from the farms and forests, to the streams and estuaries, to the oceans—to assure ourselves a continuous, reliable source of food.

Fiber

Cotton is a valuable source of natural fiber. We can substitute polyesters and nylon (which we produce by chemical synthesis), but the popularity of cotton for clothing is ample evidence of its value to us. Silk, spun by silkworms, which feed on mulberry trees, is a unique natural fiber. Wood products, from building materials (lumber) to paper, obviously come from trees. Trees support themselves and grow, like all plants, via photosynthesis. To maintain our supply of these vital products, we need healthy forests. Our agriculture and forestry industries are really managed ecosystems, from cornfields to cattle ranches to tree plantations. Agriculture and forestry can be viewed as applied ecology. Their products, too, are services derived from natural and managed ecosystems.

Clean Air

Photosynthesis is the source of the oxygen we breathe. Plants remove CO_2 and produce oxygen (O_2). Trees not only shade our homes, but they purify and renew our air. Human activities, from driving cars to breathing, increase the atmospheric concentration (and burden) of CO_2. Plants use this CO_2 as fuel and release pure oxygen. Where would we be without green plants? We would be extinct.

Clean Water

We require clean water to drink. Where does it come from? When organisms, notably decomposers, break down dead organic matter, they respire. The by-product of respiration, along with CO_2, is pure, clean water. Wetlands and riparian (streamside) forests absorb nutrients from surface and groundwater. Consequently, people have used wetlands as water purification systems for centuries. Riparian vegetated buffers trap sediments, which would otherwise run off the land (from construction or agriculture). Forested buffers provide the infrastructure of fine roots for the absorption and processing of polluting nitrates in groundwater. The mychorrhizal fungi are part of the recycling of nitrogen. They live attached to tree roots and facilitate the absorption of nutrients from the soil. Denitrifying bacteria on the root hairs intercept the nitrates in the groundwater and convert them back into free nitrogen (N_2), which is the main constituent of air.

Eutrophication results from an imbalance between production and respiration in surface waters. Excess nitrates in water may stimulate excess algal growth and lead to eutrophication. Plants also take up phosphates, the other culprit in eutrophication. Excess nutrients lead to more algal growth than the animals can eat, so the plant tissues build up, often clogging waterways. When the plants die, the decomposers (especially bacteria) have a feast. The respiration of these immense populations of bacteria can consume all the oxygen in the water, resulting in fish kills. This is exacerbated in hot weather, when the oxygen storage capacity of (warm) water is already low. So, by removing nutrients, the wetland forest or the terrestrial ecosystem guards against eutrophication of adjacent freshwaters and provides a vital service: clean water.

Healthy rivers and streams act as natural recycling centers. Decomposition is the principal energy pathway in woodland stream ecosystems. Dead organic matter is broken down into smaller and smaller particles and fed upon by a host of aquatic insect larvae and other macroinvertebrates. Salt marshes and freshwater wetlands, which border estuaries, streams, and rivers are other centers of decomposition, nutrient recycling, and water purification. Another important dynamic in streams is the cleansing of particulate organic matter out of the water. Many stream macrobenthic invertebrates like blackflies (Simuliidae), net-spinning caddisflies (Trichoptera), sponges (Porifera), and midges (Chironomidae) are filter feeders. They strain the particles out of the water, which is flowing past them. They use their appendages or small nets to collect particles (bits of detritus and living bacteria) and eat them. They provide similar services to a water purification plant at no cost to us.

The importance of streamside buffers cannot be overestimated. For example, recent legislation in North Carolina has made it illegal to cut down existing riparian forested buffers in the nutrient-sensitive Neuse River drainage. I have participated in negotiations with the timber industry to establish new rules to protect the forested buffers along streams from logging. Clear-cutting is detrimental to streams. Protecting riparian buffers is the wisest use of riparian forests since they cleanse our waters. Without riparian buffers and swamps, we would have to build extensive water purification plants. New York City just faced this issue head on. The city chose to protect the Hudson River watershed at a cost of $1 billion. The money is being used to purchase lands and conservation easements along the waterways. According to Columbia University ecologist Stuart Pimm, this will

Table 6.1
Ecosystem Services

Service Provided	*Which Processes*	*Which Organisms*
Food production		
Vegetables	Primary production	Wheat, rice, corn, soybeans
	Pollination	Bees, ants, birds
Meats	Herbivory, secondary production	Cattle, sheep, hogs, chickens
Fish	Carnivory	Tuna, flounder, cod, etc.
Fiber production		
Lumber	Primary production	Trees
Paper	Primary production	Trees
Clothing	Primary production	Cotton, hemp
	Secondary production	Wool (sheep), silk (silkworms)
Clean air	Photosynthesis	Green plants
	Nutrient cycling	Decomposers (bacteria, fungi)
		Nitrogen fixers (peas, beans)
Clean water	Nutrient cycling	Decomposers (bacteria, fungi), wetland trees
	Filtration	Invertebrate filter feeders
	Herbivory	Aquatic insects, fishes
Genetic resources		
	All	All
Natural beauty	All	All
Recreation	All	All
Climate control	All	All

save New York City the cost of building and operating water treatment plants, at a cost of over $5 billion. In summary, ecosystems provide essential services for us, and they are absolutely free. If the ecosystems were not doing these jobs, we would have to pay for them ourselves.

This discussion has given only the highlights of essential ecosystem services. See table 6.1 for the full range of services provided to humankind by ecosystems. From recreation to natural beauty, they greatly enhance our quality of life.

Ecosystem Services and Biodiversity

What is the connection between biodiversity and ecosystem services? We know that every species does a variety of jobs in the ecosystem. In most ecosystems, several species do the same job in slightly or greatly different ways. For example, all green plants photosynthesize. However, ferns do it best in the shade, and corn does it best in full sunlight. If a process is critical to the ecosystem (such as primary production), having more species guarantees that the task will be accomplished (at levels sufficient to meet the needs of the system). Ecosystems with high biodiversity are better equipped to continue functioning when stressed. An important measure of ecosystem health is thus its biodiversity, and it is the species-rich, healthy ecosystems that best provide their vital services. Since disturbance is the key to maintaining high biodiversity, disturbances are also important to ecosystem health and services.

How does this work? An abundance of species in a community increases the opportunity for a wider range of the species' traits and their ability to be represented, and it is a measure of the probability of the presence of species that have particularly critical traits. The broader the selection of species, the more likely that the key trait will be found among them and the more likely that the process gets done, and gets done well. This in turn allows for more efficient resource use in an area. Think again about primary production. Some plants specialize in each kind of light environment: from full sun (corn), to partial shade (impatiens), to full shade (ferns). If we only had one plant species, it would be either shade tolerant or sun loving but would not photosynthesize optimally in all habitats. If we have three species we might have one of each type, and each would perform at its maximum efficiency. (Note that if a plant requires shade, it is possible to give it too much light. I have personally killed several shade-loving houseplants, particularly philodendrons, by putting them in a sunny window.)

With an increasing number of species, we increase the probability that all vital traits are present in the ecosystem. Loss of biodiversity implies the loss of possibilities. Changes in species composition are likely also to alter ecosystem processes through changes in the functioning of the remaining species present. Such changes in how individual species function (or behave) can undermine the stability of ecosystem processes. Having more species available to perform a particular function is called *redundancy*. An abundance of similar functioning species provides stability of processes through redundancy and compensation. In 1997, a group of ecologists, led by Terry Chapin, wrote the following explanation. "*The more functionally similar species there are in a community—that is, the greater the diversity within a functional group—the greater will be its resilience in responding to environmental change, if those species differ in environmental responses.*" Disturbances pose far less risk to the ecosystem if the system is diverse. Biodiversity is enhanced by disturbance and also buffers the ecosystem services against interruption. So, since all ecosystems are disturbed and disturbances kill or remove some species, then, if the system is species rich, there will be more surviving species available to continue performing that particular ecosystem function. If several different plants are available in the system to photosynthesize, they can substitute one for another. This functional redundancy assures the successful continuation of the process of primary production in the face of environmental fluctuations or disturbance. I call this the "Eli Whitney Theory of Diversity and Stability." (Eli Whitney's greatest invention was not the cotton gin! He also invented the use of interchangeable parts in manufacturing.)

In abandoned farmlands (oldfields) in Minnesota, increasing plant species richness results in increased rates of nitrogen uptake. This is because different species have complementary patterns of resource acquisition.

Increased species richness increases the probability of the presence of the most productive or efficient species. If one plant gets diseased, the task of photosynthesis still continues. Furthermore, when it dies, the neighboring plants can grow larger in its stead (and in its place), thus compensating for its loss. Under changing environmental conditions, the importance of functionally similar species increases.

Species diversity contributes to the long-term stability of ecosystem functions (and, hence, services) by buffering ecosystem structure and processes against environmental fluctuations. This is a critically important value of biodiversity. There are other ways that increasing diversity can increase stability of ecosystem services: it increases the number of alternative pathways for energy or nutrient flow, thus stabilizing energy or nutrient flows among the species. High diversity can reduce the susceptibility to invasion by exotic species (species with "novel" ecosystem effects) after disturbance. Higher diversity can reduce the spread of plant pathogens by increasing the mean distance between individuals of the same species. No matter which aspect of biodiversity you focus on, higher biodiversity increases the stability and efficiency of ecosystems and their services.

To summarize, biodiversity is valuable because it contributes to the long-term maintenance of the system by buffering ecosystem structure and processes against disturbances and disasters. Species diversity is ecological insurance against disasters. Disturbances help maintain species diversity of the community, and biodiversity in turn benefits the ecosystem. Thoreau's claim that "in wildness is the preservation of the world" is not an overstatement. The world depends on natural, healthy ecosystems and all their biodiversity. Ecosystems provide valuable services to people and the biosphere. These services are so important that if the ecosystem were not doing them for us, for free, we would have to do them for ourselves—and pay for them—just to survive. Disturbances generate and maintain biodiversity, and biodiversity is essential to the healthy functioning of ecosystems. We must learn to value what ecosystems do for us and the vital and fundamental role that disturbances play in maintaining biodiversity.

Further Reading

Adams, D., and M. Carwardine. 1990. *Last Chance to See.* New York: Harmony Books.

Chapin, F. S., B. H. Walker, R. J. Hobbs, D. U. Hooper, J. H. Lawton, O. E. Sala, and D. H. Tilman. 1997. Biotic control over the functioning of ecosystems. *Science* 277: 500–504.

Connell, J. H. 1978. Diversity in tropical rainforests and coral reefs. *Science* 199: 1302–1310.

Daily, G. C., and J. S. Reichert, eds. 1997. *Nature's Services: Societal Dependence on Natural Ecosystems.* Washington, DC: Island Press.

Earle, S. 1995. *Sea Change: A Message of the Oceans.* New York: Putnum.

Chapter 7 ***Human-Caused Disturbance: All Disturbances Are Not Created Equal***

Don't it always seem to go,
You don't know what you've got 'til it's gone.
They paved Paradise and put up a parking lot.
—Joni Mitchell

Chapel Hill, North Carolina, June 2000

North Carolina calls itself the "Good Roads State." The goal of the North Carolina Department of Transportation is to bring a paved road within one mile of every residence. On the face of it, that seems admirable: it would improve access for all people to the goods and services of the state. However, what are the costs? Every road rips through the forest, leaving scars 150–300 feet (50–100 m) wide and often hundreds of miles long, eroding tons of soil into our streams and polluting our waterways. The costs must be measured not only in dollars but also in the losses of biodiversity, ecological integrity and stability, and ecosystem services. This reality is not limited to North Carolina. New roads are being built all across the country.

A common fallacy is that increasing the number and quality of roads relieves traffic congestion. Yet, as every commuter knows, the relief is short lived. Interstate 40 was completed through North Carolina in June 1990. By 1998, the traffic congestion was so bad between Chapel Hill, the Research Triangle Park (RTP), and Raleigh that morning and evening rush hour traffic snarls and delays became routine. The response from the N.C. Department of Transportation was to construct two new lanes, making I-40 four lanes in each direction from RTP to Raleigh. This increase will help for a while. Yet the truth of the matter is that better roads attract more cars, so that in a few short years we'll be back where we started.

Why should this be the case? Whether it is I-40 in North Carolina or the rain-forest highway BR-364 in Rondonia, Brazil, road construction spurs development. People want to live where transportation is good and effortless. Housing subdivisions are built where good roads make access easy. The increased population density in turn spurs the construction of shopping centers, making a positive feedback loop. More people demand more roads. More roads spur more population growth. More population brings more cars, more traffic, and the demand for new roads starts all over again.

BR-364, begun in 1980, opened up vast new tracts of previously inaccessible rain forest. This started the inexorable and vicious chain of events: clear-cutting the rain forest, burning the debris, ranching, and then abandoning the land. The road building, for all its good intentions, spurred the destruction of the rain forest. This pattern is being repeated in the year 2000 in Venezuela, as the government and the oil companies are pushing new roads into the rain forest. The road brings people, and with people comes ecological devastation, from deforestation to habitat destruction and fragmentation. The problem is global. Whether in North Carolina or Brazil, superhighways create impassable barriers to animal and population dispersal. The fragmentation of the forest threatens the very survival of hundreds of species. Is this progress?

Human versus Natural Disturbance

I have tried to make a strong case in defense of natural disturbances and their value to biodiversity, to ecosystem services, and to humanity. So, a logical question is: "If disturbances are so good, then should we just disturb everything, everywhere, all the time?" The answer is a resounding *no*! Disturbances are good for ecosystems and communities, but only up to a point. For disturbances to be effective in preserving biodiversity, an intact community must be available as a source of colonists. If everything is destroyed, there will be no organisms left to repopulate the disturbed area. To preserve biodiversity, we must preserve habitat. In this chapter I will explore the differences between human and natural disturbances.

The ideal disturbance regime (in the Intermediate Disturbance Hypothesis) is when the disturbance is neither too rare nor too common; neither too big nor too small; neither too strong nor too weak; but just right. It is at this balanced level that communities attain peak diversity. A disturbance of too great an extent or too powerful a force can be devastating and kill virtually everything. Under natural conditions, disturbances of such intensity and magnitude are rare events. The eruption of Mount Saint Helens in Oregon in 1980 covered 70 square miles (180 km^2) with ash and lava, destroying nearly all living things. Yet even with such destruction, the community is slowly recovering. Most natural disturbances are on a far smaller scale. Human-caused and natural disturbances differ in extent, magnitude, scale, and persistence. The very principles that make disturbances beneficial at natural scales can become problematic at larger scales.

Disturbance Size and Intensity

Let's take Manhattan. When the Dutch landed and first set up a trading post on Manhattan Island in 1613, it was an island of unbroken forest. We have indeed "paved paradise and put up a parking lot." Actually, we have paved thousands of parking lots and built tens of thousands of skyscrapers and apartment houses. The only seminatural remnant of the original forest community is Central Park, which itself is mowed and manicured. In Manhattan, as in other major cities, we actively maintain our grip on nature. If the cement cracks and grass pokes through, we repave it, holding the natural tendency of the community to colonize and rebound in check. People are persistent and tenacious. There is virtually no opportunity to recover the biodiversity or ecosystem services of Manhattan. It would take thousands of years to do so. That would be far longer even than the time necessary for the recovery of the community on the slopes beneath Mount Saint Helens. *We have stopped nature in our tracks.*

Humans have the capacity to disturb natural ecosystems at an unnatural scale. While people are part of nature, we can act in ways that exceed even the most powerful, most destructive forces of nature. We, ourselves, have become a force of nature. In the summer of 1997, farmers in Indonesia, as part of the traditional slash-and-burn cultivation, set massive fires. Normally, each of these fires is contained on a few acres. However, by September 1997, due to strong winds and the late arrival of the monsoon rains, the fires swept out of control. On South Sumatra, 5 million acres burned, and in East Kalimantan (formerly Borneo) another 1.25 million acres. This cataclysm dwarfed the Yellowstone fires of 1988. Smoke from the fires closed airports in Malaysia and Singapore, and aggravated respiratory diseases in people throughout the region. The death toll was 240 people. The fires and the associated smoke caused an estimated $4.4 billion worth of damage. These losses included the direct losses of agricultural production and timber and the indirect damage to ecosystem services (loss of erosion protection, storm water control, and increased atmospheric pollution).

Human disturbances are often far greater than natural disturbances. In the Indonesian fires, whole islands were ablaze. The destruction of rain forests in the Brazilian Amazon by logging companies is several orders of magnitude larger than the treefall gaps caused by storms. By 1984, people had cut down more than half the 6 million square miles of rain forests worldwide (in less than forty years). In the 1980s, the tropical rain forests of the Brazilian Amazon were being cut down at an estimated rate of 20,000 square miles (50,000 km^2) per year. The wholesale logging of the rain forest and subsequent burning of the brush and branches produce fires so immense that they can be seen from the orbiting space shuttle. Joe Connell used disturbances in rain forests to illustrate the benefits of disturbance to biodiversity. However, the gaps he envisioned were naturally produced and far smaller. At this human scale, we now find the rain-forest ecosystem threatened with total destruction. Scale matters!

Why is the size of the disturbed area so critical? Under smaller, more moderate natural fires or windthrows, the community is readily repopulated by rain-forest species, and the resulting natural system is a patchy, highly diverse rain forest. On the other hand, loggers clear-cut vast areas of forest. When modern, mechanized logging operations are finished clearing a rain forest and the trees have been removed, workers set fires to finish the clearing and burn off the remaining branches and stumps to prepare the land for cow pasture. The resulting gap is not measured in a few acres, but in hundreds of thousands of acres. The natural recolonization mechanisms of regrowth, migration, and recruitment are largely stymied. Regrowth is halted because the heat of the fires kills the roots of the rain-forest plants. Migration is limited because the area is so immense that migration from the margins of the burn is insignificant. Recruitment does proceed, but the early recruits are grasses, spread by windblown seeds. The clear-cuts and subsequent fires create a disturbance of such great magnitude that the rain-forest community is thrown into an entirely new ecological state. It will take decades or more to regrow. In fact, the denuding of the landscape results in such drastic changes in the local climate that the rain forest may never regrow. Without the forest cover, the soils warm quickly and dry out. The loggers have removed the timber, where nearly all the nutrients were stored. The fires vaporized much of the stored nutrients (in soil and litter) that were left behind. The resulting habitat is hot, dry, and nutrient depleted, incapable of growing a rain forest. The ecosystem is completely altered and its services are lost for decades, if not forever. We can find similar examples of the scale of human disturbance in every ecosystem, fundamentally altering the whole system.

The "Permanence" of Human Disturbance

Natural disturbances are typically quick and of short duration. Wayne Sousa, an ecologist, defined disturbance as "a punctuated killing." While some human-caused disturbances are short term, far more last for a very long time. The key difference is in acute versus chronic change. A tree can fall into a stream and form a temporary debris dam. That dam may last a few weeks to a year or more. The debris dam is part of the natural morphology of the river. In contrast, when engineers build a dam, the river is impounded for decades and the dam redefines and completely changes the nature of the river. The river is changed into a lake, which is essentially permanent as far as fish and insects are concerned. Lakes perform vastly different functions than rivers, being principally producers of algal biomass. Rivers are mainly consumers of organic matter through feeding and decomposition. Lakes are dominated by plankton, rivers by benthic organisms. The entire structure and function of the two ecosystems are different. Beavers dam streams and change their function, but the dams erode and break and the stream recovers. If the human engineers have done their job well, the conversion of the river into a lake is permanent.

The differences between the effects of beaver ponds and man-made reservoirs are twofold: magnitude and permanence. Let's consider magnitude or scale first. When a river is naturally disturbed, it is promptly recolonized by drifting organisms from upstream and by strong swimmers from downstream. When we insert a reservoir into the flow, the new lake forms a large, still body of water (full of predacious fish), which is an effective barrier to movement of aquatic insects. The upstream and downstream reaches of the river are often cut off from each other, so neither drift nor swimming will get the insects across the lake. The only recolonization connection that remains is the flying adults. Recall that most aquatic insects have an aerial adult stage, during which they fly, mate, and then the females lay their eggs in the water. If the lake is too large, the females will not be able to fly between one flowing reach and the other. For example, Lake Mead, formed by the Hoover Dam on the Colorado River, has an area of 233 square miles (603 km^2); it is nearly 210 miles (350 km) long—greater than the flight range of almost all aquatic insects. The dam and its reservoir transform the river, chopping it into two distinct fragments, effectively isolating each from the other and preventing recolonization from one to the other. Given the short life spans of aquatic insects (from less than one to four years), the lake will be there for many of their generations. The two reaches are really no longer even parts of the same ecosystem because migration (and gene flow) between their populations is impossible. If conditions in one portion of a fragmented ecosystem deteriorate, a population there can decline drastically. If a subpopulation is too small to maintain its genetic diversity, extinction is just around the corner.

Fragmentation of forests by land clearing and road building reduces the size of forest stands and the effective size of forest-species populations. Many forest dwellers rely on intact forest stands. "Edge species" live on the edge of the forest, thrive in the brightly lit edge, and are able to tolerate the higher temperatures and drier conditions there. Two of these are poison ivy (*Toxicodendron radicans*) and kudzu (*Pueraria lobata*, nicknamed the "vine that ate the South"); kudzu climbs and shrouds the native trees that live on the edge of the forest. However, the interior species, e.g., Red Oak (*Quercus rubra*) and Southern Red Oak (*Quercus falcata*), require the cool dark conditions of the deep woods. When people insert a super highway, a housing subdivision, or a mall into a forest, much of the forest interior is converted to edge. Roads with their long ribbon shape cut through the forest and are particularly devastating, creating edge effects on both sides for hundreds of miles. The edge effects can penetrate 90–150 feet (30–50 m) into the forest, reducing the effective population size of interior species. This is a very human type of disturbance. In nature such isolating events—like the rising of the Rocky Mountains or continental drift—are slow and rare.

When a population is too small and too isolated, a loss of genetic diversity occurs due to inbreeding, which can lead to the emergence of genetic diseases and ultimately to the extinction of the local population. This is the beginning of the loss of biodiversity with all the attendant problems. The damage to the ecosystem resulting from fragmentation is direct and severe. Natural disturbances, even big ones, are typically milder and less persistent than human disturbances.

Human Threats to Biodiversity

Human activity places severe stress on ecosystems. Many of these activities (dam and road building, land clearing, etc.) change the fundamental way the ecosystem is structured and threaten the survival of many species. Human impacts on ecosystems range from local to global problems. More species have gone extinct in the last fifty years than at any time since the mass extinction of the dinosaurs, perhaps even more than in any period of Earth's history. Land-use changes constitute the major threat to biodiversity and endanger the very existence of many species. The causes include deforestation, urbanization, desertification, and overexploitation of nature.

A recent study by Dr. David Wilcove (a conservation biologist) and others details the threats to biodiversity in the United States. They examined the causes of the threats to endangered species, and then tallied the number of species threatened by each cause. The percentages are shown in figure 7.1. Habitat destruction

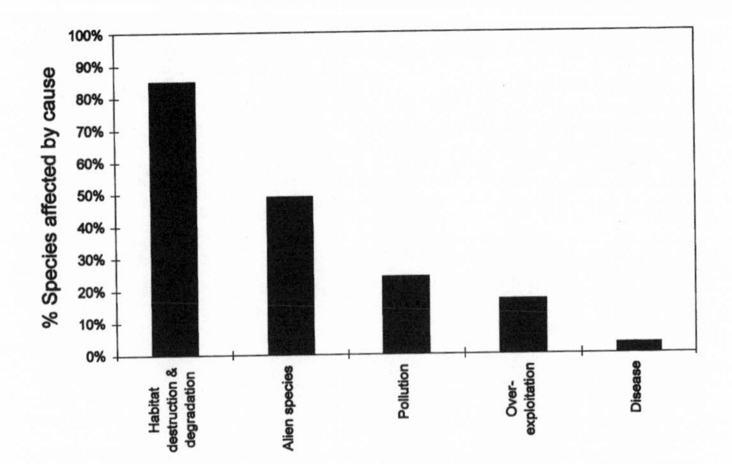

Figure 7.1 The major threats to biodiversity. Data refer to species classified as imperiled by the Nature Conservancy and to all endangered, threatened, or proposed species, subspecies, and populations protected under the Endangered Species Act. (From Wilcove et al. 1998)

and degradation is the leading cause, threatening 85% of all imperiled species. Exotic (alien) species imperil about 50 percent. Since these exotic species are often released from the competition and predation they faced in their home environment, they can grow rapidly in the new habitat, typically at the expense of native species. Not all exotic species erupt, but those that do, such as kudzu, Russian Olive, and the rodent Nutria (*Myocastor coypus*) have caused major problems for native species. Pollution threatens 24 percent of imperiled species, overexploitation 17 percent and disease a scant 3 percent. The destruction and degradation of natural habitats is therefore the worst conservation problem we have. If you want to preserve species, you must preserve habitats. Habitat fragmentation is a major threat to biodiversity. As the human population expands, we chop our remaining natural habitat into progressively smaller and smaller chunks. If this continues, the loss of more and more species is inevitable. The underlying cause of all these problems is human population growth.

Human incursions into natural habitats are occurring at the highest rates and with the most devastating impacts in the very ecosystems that have the highest rates of primary production. Recall that the most fundamental life support function of ecosystems is fixing solar energy through photosynthesis, that is, primary production. The most productive systems on Earth are also the ones most threatened by humans. These endangered ecosystems are tropical rain forests, estuaries, wetlands, and coral reefs: the very ecosystems that have the highest biodiversity.

Tropical Rain Forests

Rain forests are the most diverse ecosystems on Earth. Although they account for less than one-seventh (14 percent) of the land area of the planet, more than 50 percent of the world's species are found there. Rain forests are being cut and burned at a frightening rate. The astonishing current estimate (*State of the World 1998*, Worldwatch Institute) is that one hectare (10,000 square meters = 2.47 acres, the equivalent of 2.25 American football fields) is being cut down every minute of every day! Note that much of the high diversity of the rain forest has always been based on the natural pattern of local disturbances and low nutrients, which favor high species richness. In that patchy disturbance regime and poor soil, no single competitor could dominate. The degree of human disturbance now threatens to change the rain forest into scrub grassland and savanna, perhaps forever. The loss of rain forest biodiversity greatly limits its vital contribution of primary productivity to the planet. Furthermore, the burning of the rain forest adds tons of greenhouse gases (especially CO_2) and particulates to the atmosphere, contributing to the "greenhouse effect" and global warming. For example, the massive fires in Indonesia in the spring of 1997 blackened the skies from the Philippines to eastern Europe. Any way you look at it, rain-forest destruction is a global, man-made disaster.

Coastal Development: Estuaries and Coral Reefs

More than half of the human population, nearly three billion people, lives along the ocean coastlines of the world's oceans. Coastal development includes housing, marinas, shopping centers, industry, agriculture, tourism, and road building. The nearly unavoidable consequence of the boom in coastal development is increased runoff of sediments and pollutants. Estuaries are the most productive and one of the most diverse of all marine ecosystems. In estuaries, the effects of pollution are those of nutrient enrichment and eutrophication, often leading to fish kills (see discussion above), loss of oyster and scallop beds, and the death of sea grasses. Sedimentation and dredging (to keep shipping channels open) contribute to the problems.

Coastal pollution is a contributing factor to the destruction of coral reefs, which the most diverse of all marine ecosystems. They are threatened by direct exploitation by way of the harvesting of corals for sale and overfishing. Dredging, pollution, and sedimentation are destroying the nearshore coral reefs. Ocean warming and coastal pollution combine to further stress coral reef communities. Coral reefs respond by expelling or killing the coralline algae (zooxanthellae) that support them, a process called *coral bleaching* (since the remaining coral skeleton is pure white, with neither algae nor living coral polyps to color it). Bleaching is evidence of a dead coral reef and consequently spells the death of the highly diverse community, which was supported by the living zooxanthellae and coral. The very high biodiversity of coral reefs has also been attributed to their natural disturbance regime. However, the stresses that coral reefs are now encountering threaten their very existence.

Wetlands

Some of the most important ecosystems to people are the wetlands. They are also the most underappreciated. Wetlands include freshwater swamps (wetlands with trees), such as Okefenokee Swamp, and marshes (wetlands without trees). Marshes can be freshwater cattail marshes, sawgrass prairies like the Everglades, or salt marshes with salt-marsh cordgrass (*Spartina alterniflora*) as on the Mid-Atlantic coast of the United States. Yet, in many people's minds, wetlands are just waste places, places of no importance, or even worse, somehow threatening and evil. The dark, dank swamp is an icon of evil in literature and the cinema. Wetlands actually are any terrestrial ecosystem characterized, for a significant portion of the year, by water-saturated soils and hydrophilic (water-loving) vegetation.

Presently, wetlands are being drained and filled for housing, commercial developments, and agriculture. We have seen (chapter 6) how vital healthy wetland ecosystems are to ecosystem services ranging from nutrient retention to flood control. Under natural conditions, wetlands fluctuate in water level. The soil may be under water for several months one year but for only a few days the next. These natural fluctuations are part of the life cycle of a wetland. Many wetland species (some of which require long periods of inundation) advance and retreat with the water level and the hydroperiod (the timing and duration of inundation). This process maintains the dynamic nature of the system, keeping it patchy and diverse.

Yet, many land developers ignore the requirements of the wetland species. To them, wetlands are just nasty impediments to building houses and shopping centers and their solution is to drain them and convert them to dry, buildable land. From the swamps of eastern North Carolina to the Meadowlands of New Jersey, from the agricultural fields of Illinois to the Everglades, and so on around the world, the destruction of wetlands continues. Hundreds of thousands of acres of wetlands have been drained and filled. Once they are cut off from the natural flow of waters and the periodic inundation of the landscape, the wetlands dry out and the native species are doomed to extinction. The cost in lost biodiversity and ecosystem services is incalculable. For example, in the past, small towns have used streamside wetlands (swamps) as their principle means of sewage disposal. Today, riparian swamps absorb excess nutrients from agricultural runoff and help keep streams clean.

Spartina alterniflora salt marshes exist where few other plants can, in the intertidal zone on ocean coastlines. They encounter enormous fluctuations in salinity and temperature with the tides. At low tide, the salinity shifts toward freshwater and at high tide it goes back to saltwater. If the soil is exposed at low tide, temperatures can increase dramatically, then cool suddenly when the tide comes in. Few animals eat the live, tough cordgrass. Dead *Spartina* is an important source of detritus for nearshore fauna. The salt marsh provides food and shelter for young fishes and is the basis for significant amounts of marine productivity. Most shellfish (oysters, clams, crabs, and shrimp) spend critical juvenile stages of their lives in wetlands. They, in turn, are fed upon by fish. Salt marshes are a critical part of the entire marine food chain and ecosystem. As the bumper sticker says, "No wetlands, no seafood!"

The precise definition of wetlands is often the subject of virulent debate between regulators and developers. At issue is the U.S. policy of "no net loss of wetlands." Current policy prohibits destroying wetlands without replacing or restoring them either on site or in a "wetlands mitigation bank," a constructed or restored wetland somewhere else. The argument centers on the exact number of days when standing water is visible. The precise definition of what is and what is not a wetland can determine whether a shopping center or housing development can be built on the land. Thus, the definition of what is a wetland can mean millions of dollars to a big developer. In eastern North Carolina, in 1998–99, a six-month loophole in the regulations controlling draining and filling of wetlands resulted in a (wet)land rush as developers drained 6,000 acres (2,430 hectares) before the new regulations, which banned such activities, went into effect. This sort of cavalier and antiecological practice simply must stop. Our vital wetlands are threatened by development, by sedimentation, and by pollution. We must protect them, both for their unique flora and fauna and because they provide essential ecosystem services.

In rain forests, estuaries, and coral reefs, humans are damaging the very ecosystems we depend on for crucial life-supporting ecosystem services. We are fouling our own nest. Although I have focused here on the most sensitive, diverse, and productive ecosystems, human population growth, overexploitation, urbanization, ecosystem fragmentation, pollution, global warming, and climatic changes threaten all ecosystems, from the African savannas to the most pristine Canadian or Ukrainian lakes. We are endangering all ecosystems and hence ourselves.

Pollution: A Matter of Scale

Pollution is a shorthand, catchall term for a broad class of human disturbances that threaten biodiversity. Sedimentation—the dumping of sediments such as sand, silt, and clay—is the single greatest pollutant of streams. It is the by-product of any land disturbance—from grading to homebuilding, from commercial construction to forest or wetland clearing. It is the most widespread form of pollution affecting watershed processes worldwide. Soil runoff into rivers and streams leads to the deposition of fine particles that cover the coarser natural sediments (gravel, cobbles, and boulders) and inundate and bury high-diversity benthic communities. Sediment can blanket the bottom, build up behind debris dams, and form sandbars. Organic matter, entrained in the sediments, can exert significant biological oxygen demand and lead to anoxic conditions and fish kills. Sedimentation rapidly degrades and destroys fish and wildlife habitat.

Some sedimentation occurs naturally. For example, when floodwaters undercut a stream bank, the bank may collapse, dumping sediments into the stream. The difference between natural sedimentation and sedimentation from human land-clearing and grading activities is one of scale. Humans dump far more sediment into streams than what occurs during normal, natural processes. The time for the stream to recover from sedimentation is proportional to the amount of sediment dumped and the pattern and timing of floods. The more sediments that accumulate in a stream, then the larger the flood (and, therefore, the rarer the event) it will take to wash them out. Sediment deposition is the most critical and the most ubiquitous environmental problem in aquatic ecosystems—in lakes, streams, rivers, and estuaries.

Damming a stream to open up new fields for agriculture compounds the problem. The plowed fields typically become a new source of sedimentation into the stream. Yet, when floods are prevented by dams, the sediments stay put and the damage to streams becomes far worse. Habitat diversity is lost, and with it, biodiversity. The riffles get filled in and the stream becomes sluggish. The dissolved oxygen drops until the stream is no longer a fit habitat for fish and riffle insects. Without the dams, the added sediments would naturally get flushed downstream during floods. Fortunately, a new understanding of the essential requirement of floods for healthy stream ecosystems is emerging in the management of dams (chapter 5). By releasing large amounts of water at critical times to mimic spring snowmelt, floods, or winter storms, dam operators can aid in stream restoration.

Most natural nutrient enrichment, such as dead leaves or animals, is transient. Autumn leaves fall into a stream or a raccoon dies. Extreme nutrient additions are disturbances. Yet, these can be ameliorated by the next flood. The carcasses of spawning salmon can contribute vast amounts of nitrogen and carbon to the stream and fuel primary productivity. These nutrients could lead to eutrophication, but the next big rain will flush out most of them. This can happen with sewage spills. If they are small, the stream community can absorb the disturbance, but if they are too large, then severe eutrophication may result—and even fish kills. In contrast, human nutrient enrichment through agricultural runoff is persistent, making the nutrient addition a form of pollution. This process recurs throughout the growing season, year after year. We call it a *press disturbance,* as opposed to the more common "all at once" *pulse disturbances.* Agricultural nutrient pollution is compounded by sedimentation, and together, these disturbances can cause permanent changes in the receiving stream.

We can lessen the impact of agriculture on stream communities by requiring streamside (riparian) buffer zones along field margins. This is the goal of the new Conservation Reserve Enhancement Program of the U.S. Department of Agriculture and the North Carolina

Division of Soil and Water Conservation. The program provides the resources for North Carolina to lease 100,000 acres of stream buffers from farmers along four sensitive watersheds (the Neuse River, the Tar-Pamlico, the Chowan, and the Jordan Lake watersheds). By taking the land out of agricultural production and allowing the forest vegetation to regrow along the stream banks, sediments will be trapped by the vegetation and kept out of streams. Nutrients that might otherwise pollute the streams will be captured by plants to aid their growth and reproduction. It is a win-win situation: the farmer gets paid and the ecosystem is protected from pollution.

Other forms of pollution, such as toxic chemical spills, are not so easy to manage. Communities have evolved to exist with a wide array of natural insults; however, now hundreds of novel synthetic chemicals like kepone or PCBs (polychlorinated biphenyls) are dumped into the environment annually. Whether they are new drugs, solvents, or pesticides, they pose new challenges that natural communities have never before encountered. When a synthetic toxin is dumped into a river or on the ground and is incorporated into the sediments, the result can be permanent contamination, eradicating virtually all living things in the system. Oil spills ranging from the *Exxon Valdez* in Prince William Sound, Alaska (1989), to the Galápagos Islands (2001), to the thousands of leaking gas tanks from abandoned gas stations pose similar problems. Pollution in the form of oil spills and nuclear accidents are truly drastic forms of human disturbance, causing massive, staggering, and persistent ecological damage. Such disturbances should never be confused with the beneficial role of natural disturbances.

Ecosystem Restoration: What Works

A new area of ecological research is ecosystem restoration, a growing industry. The biggest problem in conservation is the encroachment of human habitation and development in the wildlands. It is in our own best interest to preserve these vital systems that provide us with so many essential free services. Yet all too often, the ecosystem has already been degraded and requires restoration. Ecosystem restoration is the application of our best ecological knowledge to rebuild and reconstruct natural habitats in order to restore them to their natural structure and functions, including their original biodiversity. It is ecological engineering for biodiversity and ecosystem health. Although we hate to admit it, ecologists know far too little about how to restore these systems. We can readily chronicle their demise, but restoring them to full health is very difficult. Some human disturbances are essentially irreversible. If people completely abandoned Manhattan, leaving the infrastructure intact, how many generations would it take for Nature to reclaim it?

The key to restoring degraded ecosystems is to reintroduce spatial heterogeneity and disturbances, thus rebuilding the biodiversity of the system and improving its functioning. It is trickier to reintroduce disturbances than heterogeneity.

In stream restoration, we typically start with a degraded stream that may have been channelized to make it straight for navigation, or dredged to improve drainage from croplands. In streams, restoration includes stabilizing banks, putting snag and logs back into the stream, or even recontouring the channel to replace the curves and twists that had been removed, thereby reintroducing "natural" levels of spatial heterogeneity. Often, the problem is reconnecting the river to its floodplain, especially if channelization and natural downcutting have made the channel very deep. Restoring the flow regime, including the natural frequency of floods and droughts, is more difficult and very expensive. The economical alternative is not to degrade the stream in the first place!

We are beginning to appreciate streams, floodplains, and wetlands for the diverse, productive, and interconnected vital ecosystems they are. Convincing dam operators to manage for biodiversity instead of peak power consumption is a hard sell. Nationwide, people are beginning to understand that dams are not the solution but are rather part of the problem. In fall of 1997, Quaker Mill Dam in North Carolina was demolished, reopening 230 river miles of the Neuse River watershed to migration by herring and shad. On July 1, 1999, the Edwards Dam on the Kennebec River in Maine was blown up. This reopened 17 river miles of spawning grounds to Atlantic Salmon (*Salmo salar*) and other fish. Removing dams is a quick and effective way to allow the restoration of a river and its populations. There is much pressure to decommission dams in the Columbia River basin to help salmon populations recover. Restoration of dredged and channelized streams is becoming popular with fish and wildlife agencies, with the goal of restoring the stream and its floodplain. We have a long way to go if we are to restore our nation's rivers to full ecological function and biodiversity. Wetland restoration is becoming big business, since U.S. law does permit some dredging and filling of wetlands so long as equal amounts of wetlands are restored.

The most common form of ecosystem restoration is reforestation of cut-over lands. Tree planting in the timber industry is very common, and forest restoration has its own industry. We have timber companies with vast land holdings that regularly, and very professionally, clear-cut the forest, then bushhog the underbrush, replant trees, and regrow great stands of trees. Are these tree plantations really forests? The trees are typically single species plantings of Loblolly Pines (*Pinus taeda*), other pines, or sometimes *Eucalyptus* species. However, there is something fundamentally wrong with these "forests": they are too orderly, too simple; they are arranged in perfectly symmetrical rows. When I walk in a pine plantation, it feels to me like something too perfect, too precise, too mechanical and unnatural. The single-species plantations have little in common with natural forests. There is no tree diversity and little animal diversity, so this "forest" cannot survive except by the hand of man—in fact, I "cannot see the forest for the trees." Reforestation to restore natural forest structure and biodiversity is still in its infancy.

The National Park Service and the National Forest Service, heeding the lessons of the Yellowstone fires, have begun a program of prescribed burns to keep the forest ecosystems healthy and vigorous. Yet, they must not rush ahead in their eagerness to do the right thing. It is important to design the burns to be as patchy as natural forest fires. I am concerned that the prescribed burns will be too small, too cool, in which case these fires will be too little, too late, to preserve the vigor and biodiversity of our forests.

Human-caused disturbances often act to homogenize the ecosystem, thus eliminating the patchiness that natural disturbances produce and react to. Remember that the interaction between patchiness and disturbance is what yields the highest biodiversity, and the key step in restoration is to remove the barriers to natural functioning—I have great faith in the recuperative powers of natural ecosystems. A clear-cut landscape, left alone, will eventually regrow an entire, diverse, natural forest. Remove the dam and a stream will heal itself. If we just stop manipulating these impacted systems, the natural sequence of disturbance and succession will, in time, return the ecosystem to its natural state.

Human Disturbances Are Too Extreme for Natural Ecosystems

Natural ecosystems can absorb a lot of change. Natural disturbances are part of the life of every community, yet there are limits. At some size or magnitude of disturbance, we can throw a community into an entirely new state as, for example, a dam changes a free-flowing river into a lake, or a wetland is drained and filled to make dry land for construction. At some point, an ecosystem becomes so altered that it becomes something entirely different. Unfortunately, this new system is almost always less diverse than the original, natural ecosystem and has impaired ecosystem services. I know of no case in which human-induced disturbance of the environment has enhanced biodiversity or ecosystem services. We thus need to control and minimize these human-caused disturbances. When contractors bulldoze and level a forest and then construct a subdivision, that change is essentially permanent; the forest ecosystem, with all its biodiversity and attendant ecosystem services, is gone forever. When developers drain a wetland and put up a shopping center, the wondrous diversity of the wetland is destroyed and its ecosystem services are lost. Sadly, such scenarios are being replayed daily across the United States and around the world. All of us know about a favorite stream that is now a lake or a patch of woods where houses stand. The singer Joni Mitchell had it right: we "pave paradise and put up a parking lot."

Further Reading

Eldredge, N. 1998. *Life in the Balance.* Princeton, NJ: Princeton University Press.

Harris, L. D. 1984. *The Fragmented Forest: Island Biogeographic Theory and the Preservation of Biotic Diversity.* Chicago: University of Chicago Press.

Quammen, D. *The Song of the Dodo: Island Biogeography in the Age of Extinctions.* New York: Scribner's.

Wilcove, D. S. 2000. *The Condor's Shadow: The Loss and Recovery of Wildlife in America.* Foreword by E. O. Wilson. New York: Anchor.

Wilcove, D. S., D. Rothstein, J. A. Dubow, A. Phillips, and E. Losos. 1998. Quantifying threats to imperiled species in the United States. *BioScience* 48 (8): 607–615.

Chapter 8 ***Toward an Ecological Worldview***

The woods are lovely, dark and deep.
But I have promises to keep,
And miles to go before I sleep.

—Robert Frost (1874–1963), "Stopping by Woods on a Snowy Evening"

January 1998, South Florida

The Everglades are a scene of endless delights. As you approach a pond, a snakelike creature raises up from the water. It is an Anhinga, a diving water bird, fishing for its midday meal. Soon, the amazing diversity of the waterfowl becomes apparent. A Wood Stork rises majestically from the sea of pale yellow-green grass and flies into the haze. There are Great Blue Herons and Little Blue Herons, Tricolored Herons and Green Herons, and Yellow-crowned and Black-crowned Night Herons. There are dozens of species of ducks, from the Common Mallard to Mergansers and the rarely seen Gadwall. There is the chubby, vibrant purple Gallinule, resplendent in purple, blue, green, and white. Above all of them soar the predators: Bald Eagles, Ospreys, and Snail Kites. All of this diversity is just in the freshwater ponds. Move to the bay shores and there is even more.

As you approach the patches of woods, you see a great diversity there, too. These are the hardwood hammocks (stands), raised slightly above the surrounding sawgrass prairie. Here are the Live Oaks, the Florida Royal Palm, and the Cabbage Palm; the Gumbo Limbo tree and the West Indian mahogany. Intertwined with these are the Strangler Figs, wrapping themselves around the trunks of other trees, climbing toward the light and killing their supporters in the process. All around are vines and creepers. The animals are distinctive too, most notably the Florida tree snails, with fifty or so distinct varieties, with patterns of pink, yellow, brown, and green. Common in the Caribbean, they have successfully invaded the Everglades. There are other distinctive woodlands, too. There are cypress heads, pine savannas, and mangrove swamps, each with its own distinctive flora and fauna. This bounteous diversity of life in the Everglades is what makes it such an extraordinary and special place.

The Everglades: An Ecosystem Shaped and Maintained by Disturbance

Spatial Heterogeneity and Diversity in the Everglades

The popular book *The Everglades: River of Grass* (1947) was aptly titled by its author, Marjory Stoneham Douglas. Native Americans, the Miccosukees, had named the Everglades "grassy water." Indeed, much of that region is a wetland, a sawgrass prairie dominated by *Cladium jamaicense.* However, it is the extreme diversity of community and ecosystem types in the Everglades that provides us a case in point, pulling together the various themes of this book. The Everglades contain coastal saline flats and prairies, patches of cypress forest (called cypress heads or cypress domes), salt marshes, freshwater marshes, hardwood hammocks, mangrove swamps, open pine savannas, sawgrass prairies, and tidal creeks. What accounts for and maintains this extreme complexity of community types?

The water, which ultimately forms the Everglades, begins in the Kissimmee River, which flows into Lake Okeechobee. Before the 1960s, Lake Okeechobee would fill up during the rainy season, then overflow its southern shoreline. This produced a vast sheet of water, which flowed southward, forming the Everglades. The Everglades once covered nearly all of South Florida. Unlike a river that is confined within its banks, the Everglades is bankless river, spreading broad and shallow across the entire flat landscape. The resultant flow pattern is called *sheet flow.* In the Everglades, the entire system is analogous to a vast river floodplain, with a scarcely visible, undefined channel. Water from Lake Okeechobee is constantly flowing across the Everglades, but so slowly that you can barely detect the flow. It is this broad, shallow, north to south flow that has created the river of grass. Disruptions to this flow pattern have turned the Everglades into an endangered ecosystem.

The Everglades' spatial heterogeneity is due to several characteristics of the flow of water. In addition to the directional flow, there is variation in the salinity of the waters, from the freshwater sources in the north (from Lake Okeechobee) to the saline waters of Florida Bay in the south. In addition, there is variation in the depth of the water and hydroperiod (the number of days or months that water covers the soils). Depth and hydroperiod are of major importance to the local spatial heterogeneity of the Everglades. The slight rise and fall of the land surface is the key to determining water depth. Where the soil dips and the water collects and gets a little deeper (more than 8 inches; 20 cm), we have a freshwater pond. Where the water is very shallow, we have a sawgrass meadow. Where the land rises 3 feet (1m) or sometimes even just a few inches, above the water level, then trees appear. Bay heads, dominated by Red Bay (*Persia borbonia*) or Sweet Bay (*Magnolia virginiana*), occur where a soil mound occurs. This can be the long-term result of alligators digging nests in the soil, creating mounds of earth around the nest, ultimately becoming sites for bay heads. Where the water stands for a long period due to a depression, you will find a Cypress head, with species of *Taxodium.*

Because all species are adapted to specific habitats, the diversity of habitats dictates the biodiversity of the Everglades, too. Here, the microhabitat diversity is created by the delicate interplay of topography, water depth, salinity, and flow rate. These in turn can influence water temperature and water chemistry. What is so striking about the Everglades is the tremendous impact of what superficially appear to be minute changes in the habitat. Inland, a rise in elevation of just 8 inches (20 cm) can raise a hardwood hammock community from the surrounding sawgrass prairie. At the shore of Florida Bay, the same change will generate a mangrove swamp from the surrounding salt marsh.

Disturbance and Diversity in the Everglades

In addition to the underlying spatial heterogeneity, there is yet another major factor in the maintenance of all this diversity: disturbance. The Everglades is an integrated mosaic of communities, created and maintained by hurricanes, floods, drought, and fire—a perfect illustration of the interconnectedness of spatial heterogeneity, disturbance, and biodiversity.

Fire is very important in the Everglades. South Florida is seasonally dry. The wet season is May to October, when 75 percent of the annual rainfall of 50–60 inches (127–152 cm/year) occurs. The winter-spring dry season follows. Lightning-caused ignitions of fire occur during the summer thunderstorms. (Human carelessness, the other major source of fires, is chronic, not seasonal.) These naturally caused wet-season fires typically don't consume either soil or roots and cause only temporary changes in the Everglades landscape. They rejuvenate the sawgrass prairies by returning the nutrients to the soil as ash. The pinelands are dependent on fire disturbance for their very existence, making them a "fire-climax" community. Slash Pine (*Pinus elliottii*) is dependent on fire because of its serotinous cones, which open when heated to release their seeds. Protected by a dense resin layer, it is highly resistant to fire and survives to grow again. The small Saw Palmetto (*Serona repens*) and the taller Cabbage Palm (*Sabal palmetto*) are both also highly resistant to fire. Fire suppression rather than fire is the real danger to this community. As in the Longleaf Pine ecosystem, fire suppression allows the development of a hardwood understory, whose seedlings and saplings are easily killed by fire. Slash Pines need the bare soils that result from a fire to germinate.

The absence of fire allows the litter and shade to increase, favoring the hardwoods. Two to three decades without fire will allow them to overgrow the low-stature pine trees, replace the pinewoods with a tropical hardwood hammock. Without fire we would lose much of the character, diversity, and ecosystem services of the Everglades.

The interaction of drought and fire is another source of heterogeneity and diversity. The peat soils of the Everglades are rich in organic matter and can be cut out and dried to use as fuel. Under drought conditions, the peaty soils become highly flammable. Dry-season fires (and even early wet-season fires, which begin before the ground is saturated with water) can be transforming events. If the peat burns, the level of the soil is lowered, permitting the local pooling of water. These pools, in turn, change the entire community. Lowering the soil surface increases the hydroperiod and favors the development of marshes. The tropical hardwood hammocks can survive surface wet-season fires, but they are killed by peat fires. Not only are the resident trees killed, but the resulting change in hydroperiod means that the hardwood hammock community cannot regrow there. The fire both creates and responds to the heterogeneity of the Everglades' environment, again demonstrating the interaction of disturbance and heterogeneity.

Besides fire, the other major disturbances are floods and droughts. South Florida is nearly flat, so water management is one of the key issues here. Because water moves slowly, drainage has been a priority for agriculture and development. The first major drainage canals, intended to prevent the flooding from Lake Okeechobee in the wet season, were built in 1917. This was the beginning of a massive system of canals and levees that would transform South Florida—changes that helped stimulate the tremendous growth in the human population there. Population growth required movement of water supplies to the population centers of Miami and Fort Lauderdale. Agriculture, particularly sugarcane farming, became very important in the area directly south of Lake Okeechobee and north of the Everglades. Remember that floods from Lake Okeechobee are the lifeblood of the Everglades. All these diverse interests (cities, sugarcane, and the Everglades) were demanding and competing for water.

In the end, the Everglades were the big loser. Deprived of its natural flood cycle and hydrology, the entire ecosystem has become threatened. The ancestral Everglades was more than twice the size of the present Everglades National Park. Florida Bay, the huge estuary that takes its freshwater supply from the Everglades, became progressively more saline, threatening the nursery areas for fish and shellfish species. Mangrove forests, the intertidal swamps that live at the interface between fresh and salt water, are also threatened. Red (*Rhizophora mangle*) and Black (*Avicennia germinans*) Mangroves can survive brief periods of hypersalinity in Florida Bay (during the dry season), but White Mangrove (*Lagucularia racemosa*) cannot. Red and Black Mangroves live on the shore of the bay, while White Mangrove is somewhat higher and farther inland. Groves of Buttonwood (*Conocarpus erecta*) grow on higher and drier sites, which are less likely to flood. Thus, there is a progression of tree species landward from the bay, and any change in the hydrology causes dramatic changes in the vegetation, and in the wildlife, it supports.

Now, consider hurricanes, a major form of disturbance in the Everglades. Coastal Florida gets hit by a hurricane about every ten years. A major one was Hurricane Andrew in August 1992, which devastated the Everglades communities of Homestead and Florida City, destroying over 80,000 homes along a 25-mile swath. Hurricane Georges struck Florida hard in 1998. A hurricane is a particularly devastating disturbance because its high-velocity, rotating winds bring heavy rains, huge and powerful waves, and the unique damage of storm surge. Storm surge occurs when the water, pushed by the winds, piles up against the land. Additional surface damage occurs when water that has built up in the bays behind barrier islands (as on the Outer Banks of North Carolina) as well as on the land rushes back to the sea after the storm passes.

In Florida Bay, storm surges carry large amounts of marl mud, rich in limestone (calcium carbonate, $CaCO_3$). The marl becomes trapped among the tangled roots of the Red and Black mangroves and the resulting mound raises the soil surface, suffocating these sensitive trees and killing them. The result is an ideal habitat for the establishment of buttonwood forests, as well as Black Rush meadows and Saltwort marshes. Saltwort, *Batis maritima*, is a shade-intolerant species whose success is fostered by the death of the mangroves. Many species exist only because of the openings created by the hurricanes (Sea Daisy, Glasswort, Sea Purslane, and so on). Devastating as they are to some species, hurricane disturbances thus permit the coexistence of many other species, bolstering and maintaining high biodiversity.

Now, you can see clearly that the rich mosaic of habitats and biodiversity of the Everglades is the product of the interaction between heterogeneity and disturbance. The underlying heterogeneity is due to fires, floods, and hurricanes, and the great range of their severity. All of these factors combine to create and maintain the wondrous biodiversity of the Everglades.

The Implications for Management

The South Florida Water Management District is attempting to integrate ecology and public policy. Encroachment by agriculture, the expanding populations of the cities, and fire suppression by the National Park Service have all contributed to the Everglades' decline and loss of biodiversity. The single biggest problem has been diversion of water from the Everglades, disrupting the amount and timing of flows (including periodic floods and droughts) and altering the ecosystem. One of the largest, most expensive ecosystem restoration projects in history is now underway. The details of the scientific background and the restoration plans are beyond the scope of this book (but see the suggested readings at the end of this chapter). The South Florida Water Management District has the unenviable task of trying to balance the competing water demands of urban centers, agriculture, and the Everglades ecosystem. The effort to restore the Everglades and Florida Bay is beginning to undo the damage inflicted by five decades of water diversions, flood control, and channelization. The tasks include reinstating the ancestral hydrologic regime including floods, through the reintroduction of the seasonal variability of flow, and more water to reach the Everglades. (Droughts will continue to occur naturally independent of what humans may do.)

We have seen that fire is a common feature of the Everglades and helps to create the mosaic of habitats, which in turn fosters the Everglades' diversity. Suppressing fire to protect property is in conflict with the essential role of fire in maintaining the Everglades' patchiness and biodiversity. All these changes in water and fire management must depend on the concessions made by the competing interests of the people of South Florida. The very survival of the Everglades ecosystem is at stake.

Toward an Ecological Worldview

For too long, people have viewed themselves as apart from nature, and we need to recognize that we are a part of nature. We cannot separate ourselves from our environment and live in a fully artificial world. We are absolutely dependent on healthy ecosystems for a range of ecosystem services, from a breathable atmosphere to clean fresh water to drink (see chapter 6). Sustaining these essential ecosystem services requires healthy soils to support our food production and vital, vigorous wetlands to clean our wastes and produce our seafood. We must fully recognize our dependence on our ecosystems and then act to protect and preserve them.

To achieve these ends, we must begin by reevaluating how ecosystems work and how we can work with them. Our deep-rooted fear of fires and floods must yield to an understanding and appreciation of disturbances and we must plan accordingly. The elaborate environmental management system we have created is grounded in an archaic, misguided belief that disturbances are evil and that nature must be controlled. Our history of attempting to control—indeed, to dominate—nature has led to many of the problems we now face.

We must develop an ecological worldview, a new perspective in which we will see ourselves living in harmony with nature. We need to learn to move with the rhythm of the tides, to appreciate not only nature's regular patterns such as the changing of the seasons but also the irregular ones—the untidy, dangerous occurrences: the disturbances and natural disasters. We must remember how communities and ecosystems really work. Fires and floods, hurricanes and droughts are part of the natural fabric of nature. "The nature of nature is change." We must learn to accept and appreciate nature's discords in addition to her harmonies. We need to learn to integrate ourselves into the natural landscape and live with the dynamics of nature. We must accept that disturbances are a natural, vital part of life, despite their destructiveness. High biodiversity depends on the dynamic, changing nature of communities and ecosystems, and ecosystem services, from clean air and water, to food and wildlife, are dependent on biodiversity. Biodiversity is our ecological insurance for a secure, sustainable future.

In our drive to keep nature under control, we are harming ecosystems and ourselves. From this new ecological perspective, our view of nature must be one of stewardship, not dominion; partnership, not control. We need to understand ecology and see that ecosystems get our respect and protection. In a real way, we must protect them from ourselves. We need to get out of nature's way.

Signs of Change in Attitudes toward Disturbances

As we embrace an ecological worldview with a strong environmental ethic, the resulting changes will permeate our society. When we incorporate the reality of natural disturbances and the importance and value of biodiversity into our thinking, we will change the way we plan our communities, manage nature, and behave in our personal lives when dealing with the environment. Though we often feel powerless to slow the destructive momentum of our society, signals of change are coming from our federal, state, and local governments. Let us examine some of the changes in policy and environmental management that give us reason to hope for a truly sustainable future.

Changing Federal Flood Insurance

Since now we know that floods are inevitable, we need to change where and how we build our homes. We need to prevent property losses to floods, not just insure against them. This action will take a radical policy shift, though these new ideas may be on their way to being accepted. On November 10, 1998, James Lee Witt, the director of the Federal Emergency Management Agency (FEMA) took a first step in this direction. He proposed that flood insurance should be denied to homeowners who have filed two or more claims that total more than the value of their home and who have also refused to elevate their homes or accept a federal buy-out. He further proposed that flood insurance should no longer be subsidized but sold at fair market rates, dramatically increasing the cost to policyholders. On June 27, 2000, FEMA issued a report on coastal erosion showing that one in four homes within 500 feet of a coastline are expected to be lost to erosion in the next sixty years. The study suggests that the National Flood Insurance Program should be adjusted so that homeowners with high erosion risks would pay more, as much as twice the current rate.

The current federal flood insurance encourages people to build, rebuild, and live in flood-prone areas by shifting the financial risk from the homeowners to the government. The federal government first became involved in flood insurance because the reality-based rates for flood insurance by private insurance companies had skyrocketed. Federal flood insurance insures riverfront, oceanfront, and coastal property. Since so many of these houses are second or recreational homes, this federal program has been criticized as a subsidy for the richest Americans. If these proposed policy changes survive the attacks of those who have profited from the current insurance program, we will have taken a giant step in the right direction. There really are recalcitrant people who rebuild in the same hazard zone over and over again. The American people have been paying long enough for often affluent people to have an ocean or river view, regardless of the risk to that property. In 1998, in a statement of great candor, James Witt said, "People need to accept the responsibility and consequences of their choice to live in high-risk areas. We know that there will always be another disaster. Hurricanes, tornadoes, droughts, earthquakes, fires, and floods won't stop coming." We may not be able to prevent storms and their resultant floods, but we can get out of their way.

Changing National Forest Management

For nearly all of the history of the U.S. Forest Service, the principle goal has been to supply timber to the forest products industry. This policy has included (1) clear-cutting of large tracts of national forests; (2) building of expensive logging roads or the timber trucks into the forests, at taxpayer expense; and (3) leasing timber lands for harvest to the timber companies at below their real cost. Below-cost timber leases are a direct subsidy by the American people to the forest products industry. Through the Forest Service, we build the roads and mark the tracts while the timber companies make the profits. The U.S. Forest Service is supposed to be the regulator of the timber industry and preserver of our national forests, not the provider of wood pulp and lumber to the forest products industry.

Forests are far more than tree farms for lumber. They provide habitat for wildlife and foster biodiversity with all its attendant values. They buffer the catchments for our clean water and help purify it, as over 80 percent of the headwaters of streams in the United States start in national forests. Though trees can be grown on private lands—on tree farms—there are no substitute watersheds except those in our national forests. Finally, national forests provide recreation, having three times as many visitors as the national parks (865,000,000 visitor days in 1996).

In 1998, Michael Dombeck, the director of the National Forest Service, instituted a moratorium on new road building in roadless areas of the national forests. Dombeck was trained as a fisheries biologist and understands the complex linkages between forests, watersheds, and water quality. The vast network of logging roads (380,000 miles of roads, one-half of them in great disrepair) cause sedimentation and soil erosion in our national forests' rivers and streams, degrading the water quality, impairing the habitat for fish and macroinvertebrates, and limiting the ecosystem functions and services of streams. Stopping the construction of more roads will help prevent erosion and protect water quality and stream ecosystems. In October 1999, President Clinton proposed banning all road building in roadless areas of the national forests. On January 9,

2001, Dombeck, banned all logging of old-growth timber on public lands. These are two of the most important changes in the history of United States environmental policy.

With clear-cutting, all the trees in an area are cut down at once. It is an efficient method for the loggers but devastating to the ecosystem and its wildlife. The Forest Service has already reduced clear-cutting by 84 percent from its peak annual acreage. Senator Robert Torricelli of New Jersey introduced a bill (S. 1368) in Congress in 1999 called "The Act to Save America's Forests," whose passage would ban clear-cutting in the future. Such a bill would have been unthinkable only a few years ago, and gives us hope for a future ban on all clear-cutting in the national forests. Under the new vision introduced in the bill, we can expect real multiple uses that will include logging, wilderness preservation, recreation, and protection of waters and watersheds.

Fire Policy in the National Parks

Before the Yellowstone fires of 1988, the National Park Service (NPS) had already begun to recognize the vital role that fire plays in the health of forests and in the maintenance of biodiversity. Fire suppression had been a well-established policy for nearly a century. Smokey Bear still glowers down from billboards, denouncing fires. Old ways die hard. Presently, the NPS is struggling to overcome resistance to a "Let it burn" policy, both from the public (which was raised with Smokey Bear) and from within the Park Service itself. The Park Service walks the tightrope in its mission of multiple use in all of its efforts. It must balance wilderness protection, protecting geological features, protecting biodiversity, all while running an outdoor nature museum for the nation. As national park visitation and public demand for outdoor recreation continue to increase, the conflict between public facilities and wildfire will only get worse. The conflicting pressures on the NPS are tremendous. On the one hand, ecologists and conservationists want it to allow fires to run their course, and on the other hand, tourist boards and hotel and other park facilities operators want them put out fast. Many of the ecosystems that the NPS is sworn to protect actually require fire to survive, so for the NPS to do its job, the public must understand the importance of fire.

During my 1998 visit to Yellowstone National Park, I was impressed by the Park Service's films and exhibits that emphasized the revitalizing nature and ecological benefits of forest fires. This was a genuine public education campaign. The NPS's goal (like mine) is to educate the public about the essential role of natural disturbances in ecosystems. For example, we have already witnessed that the Yellowstone fires did not destroy Yellowstone. Beneath the charred remains of the old Lodgepole Pine forest a new, healthy pine forest is already growing. Yellowstone is being re-created: it is fresh, healthy, and vital once again. The Yellowstone experience needs to be translated to all the other parks with our support. Learning the lessons of the Yellowstone fires will move us in the right direction, towards healthy ecosystems.

DECOMMISSIONING DAMS

Hundreds of hydroelectric power-supply dams are up for relicensing in the next five years. Now is the time to evaluate these dams and ask whether the power they generate is sufficient to offset their environmental and ecological impacts. While it is easy to calculate the value of the electricity they generate, now is the time also to evaluate their ecological costs, including the loss of wetlands and free-flowing rivers and all the associated ecosystem services they provide.

I am not opposed to hydropower; actually, it is one of my favorite power sources. I have had a special love for water wheels and gristmills since my childhood, and hydropower is a clean source of energy. In some situations, reservoirs are highly beneficial for fishing and recreation. However, in other situations, the damage to native species and the ecosystem outweighs the benefits (recall the negative impacts of dams on the Columbia and Snake Rivers on Pacific Salmon discussed in chapter 5). When we factor in biodiversity and ecosystem services, the cost-benefit analysis can change radically. Many dams do not generate enough hydropower to justify their continued existence. If we are responsive to the needs of the ecosystem, then decommissioning some of these dams (tearing them down, or blowing them up) may be the best public policy for the long term.

Flood control dams involve similar issues. Because these dams are designed explicitly to prevent downstream flooding, they do considerable damage to the rivers and their floodplain wetlands by disrupting the natural flow patterns and dewatering rivers for long periods of time. As in the Grand Canyon, they limit the flow of water and deprive the rivers of their revitalizing floods. Understanding the essential role of flooding for stream ecosystem health is a first step toward factoring in flood disturbance on the benefit side of the cost-benefit analysis. All of these decisions about the fate of dams will require balancing the needs of people with the needs of the rivers, whcih may prove to be one and the same. The proposed changes in federal flood insurance will help us move toward healthier rivers and streams and reduced property damage. Local governments and zoning boards have a role to play, too. With careful, intelligent floodplain zoning we can prevent construction in high-hazard zones. Once we move our houses out of the floodplain, we can return rivers to their natural, free-flowing state, and everybody wins.

Kissimmee River Restoration and the Everglades Program

I opened this chapter by describing the richness and complexity of the Florida Everglades, and the threats to their survival. The principal threat to the Everglades is the diversion of freshwater through a series of canals that divert water to the Atlantic Ocean. Beginning in 1962, the Kissimmee River was straightened and channelized from a 277 mile (166 km) meandering river into a 150 mile (90 km) trench, 33 feet (10 m) deep and 333 feet (100 m wide). At long last, a major effort has been mounted to restore this river to much of its original meandering course and variable depth. It is the largest and most expensive restoration project in U.S. history—and perhaps in the world, involving cooperation from federal, state, and local governments. Dozens of governmental agencies, including the Army Corps of Engineers, the U.S. Fish and Wildlife Service, the Florida Department of Environmental Protection, and the South Florida Water Management District, are involved, as are many nongovernmental organizations and private citizens. The most dramatic part of the plan will be to fill in some 56 miles (35 km) of the massive canals that were dug between 1962 and 1971 to provide flood control to central and southern Florida. The Kissimmee River Restoration Project will restore 27,000 acres (11,000 ha) of damaged wetlands, or about 85 percent of the total impaired area. It will return free-flowing water to remnant river channels, eventually restoring the stream invertebrate fauna. These benthic (bottom-dwelling) animals are a vital link in the food chain of fish, wading birds, and waterfowl.

Will the plan work? There are reassuring data from the Kissimmee River Demonstration Project (1984–90), that suggest it will. When three weirs (partial dams) diverted flow back into remnant river channels, the restoration of flow for only forty days resulted in natural (pre-channelization) densities of stream organisms. The number of fish increased as well. By the end of the demonstration project, wading-bird densities had doubled from the 1978–80 survey. In time, the natural mix of wetland plants will recolonize and thrive. So, restoring flows and floods can, indeed, restore streams and wetlands.

What comes next, and how will it help save the Everglades? In May 1994 the Florida legislature passed the Everglades Forever Act. The plan is to fill in many of the water diversion canals, which for decades have carried the flow of freshwater eastward into the Atlantic Ocean instead of into Lake Okeechobee. The restoration of seminatural flows in the Kissimmee River will restore the flood cycle in Lake Okeechobee as well, in turn recreating the sheet flow, and revitalizing the hydrologic regime of the Everglades, reversing the decline of the sawgrass prairies—All of this is an excellent demonstration of ecological interconnectedness. Water treatment for nutrient removal will help return the Everglades and consequently Florida Bay to their natural low-nutrient status, reversing the process of eutrophication in Florida Bay. The National Park Service is also utilizing controlled burns to return a modified fire regime to Everglades National Park and its upstream neighbor, Big Cypress Swamp. All of these efforts are designed to return the South Florida ecosystem to its natural state, through the restoration of the natural disturbance regime of floods and fires. If given enough public support, this plan will work.

Sustainable Living Means Living with Nature

Ecosystems are highly resilient. We are learning how to restore damaged ecosystems to full health and allow them to resume their vital ecosystem services. It's really not so hard. The steps are simple. First we stop manipulating and polluting them. Next, we should leave them alone, allowing the full range of disturbances to take their course. To do this we must first accept the reality, and the necessity, of natural disturbances. Then, to minimize the risk to life and property, we must get out of their way. Conceptually this is simple, but in practical terms it is harder. Many of us will need to change our attitudes and expectations about land use. We must use all the tools available to us. Zoning laws must be changed and financial incentives and penalties must be developed to move people out of hazardous areas. Ecologically sound construction techniques must be adopted so our structures will withstand the disturbances that will come. As we seek sustainable economic development based on this new ethic, we must balance human needs with the needs of the ecosystem. If we fully understand the need for these changes, then we will have the wisdom to see that, by living in harmony with our ecosystems, everybody wins.

We must build our society around this ecological worldview, which values disturbances for the good they do for the environment, the ecosystem, and ultimately for us. Then we will work to protect the world's biodiversity, both for the sake of the endangered species, and for our own enlightened self-interest. Conservation and sustainable development are not just idle contemporary catch phrases. They represent our only choice for a livable world, for our children and ourselves. We have miles to go before we sleep.

Further Reading

Davis, S. M., and J.C.D. Ogden, eds. 1994. *Everglades: The Ecosystem and Its Restoration.* Boca Raton, FL: St. Lucie Press.

Douglas, M. S. 1947. *The Everglades: River of Grass.* Sarasota, FL: Pineapple Press.

Koebel, J. W., Jr. 1995. An historical perspective on the Kissimmee River Restoration Project. *Restoration Ecology* 3: 149–159.

McNeill, J. R. 2000. *Something New under the Sun: An Environmental History of the Twentieth-Century World.* New York: Norton.

Terborgh, J. 1999. *Requiem for Nature.* Washington, DC: Shearwater Books.

Epilogue ***Living with Disturbances***

It is absolutely impossible to transcend the laws of nature.

—Karl Marx, letter of 1868

Every thousand-mile journey begins with a first step.

—Chinese proverb

The Necessity of Disturbance

Disturbances are essential to maintaining biodiversity, and biodiversity is essential for the healthy functioning of ecosystems and the stability and efficiency of ecosystem services. Once we understand ecosystems and how they work, we need the means to achieve them and to sustain their biodiversity and health. We have learned that disturbances are natural, and that fires, floods, and hurricanes are bound to happen as part of natural cycles. Let's get smart. Let's try to get out of their way. Here are a few ideas on how to achieve these goals.

Disturbances and Development: Ecologically Based Zoning

Ecologically based zoning is an excellent place to start on our path to restoring well-functioning ecosystems. Land development must be ecologically sound and controlled to minimize the risk to life and property. Those allowed to build in disturbance-prone zones will lose their buildings in time. Freedom of choice and free exercise of property rights must therefore be tempered by risk aversion and sensitivity to ecological realities. City and regional planners in all areas know what the ecological risks are in their communities and where the risks are greatest. We need to use that information to regulate development in flood, fire, storm, and erosion-prone areas, which often are home to unique species and communities. In protecting our homes and ourselves from natural disturbances, we are also protecting our ecosystems. In building structures, our goal should be long-term ecological compatibility. Below are some zoning guidelines for minimizing losses due to inevitable natural disturbances.

Coastal Development: Don't Build on the Beach!

Beachfront property is highly desirable. Along the barrier islands and the Outer Banks of North Carolina, Georgia, and Virginia this property is the most expensive. It offers the best view of the beach. Simultaneously, it has the highest likelihood of being destroyed and washed away in a hurricane. A major storm will not only tear the roofs off houses, and the houses off their foundations, but it can and will reclaim the beach itself for the ocean. Whole dunes and the houses built on them are commonly washed away in storms. Wallace Kaufman and Orrin Pilkey, in their excellent book *The Beaches Are Moving*, explain that the natural state of barrier islands is to erode at one end and reform at the other. This is an ongoing process, accelerated during storms. The movement of barrier islands is absolutely normal, and we know the risks. The only way to solve the problem of beach erosion is not to build there in the first place. Let's just say "no" to building on beaches.

Stop Development in Floodplains

Rivers *will* flood, and there is no mystery, no surprise about this. If you build a home or a business in a river's floodplain, it is bound to get flooded and perhaps destroyed. Many communities now have zoning ordinances that rely on estimates of the "100 Year Floodplain," the level that the extreme event—the one flood in one hundred years—will reach. The assumption is that building outside of that zone is safe. Flood frequency is increasing in urbanized areas because the storm waters run off the paved and built-up surfaces. So, what was once the 100-year flood mark is now exceeded far more often. These measures need to be reevaluated. Recall that floodplains and their wetlands are vital ecosystems, providing many ecosystem services to us. When we build there, we compromise the ability of the wetland ecosystem to function.

Insurance companies (and the U.S. government's flood insurance program) need to set their rates according to ecological reality. If the ecological risk is high because the development is in a high-hazard zone, then the owner should be required to pay a very high—even prohibitive—insurance premium based on that risk. Some progress is being made on this issue. After the Great Mississippi Flood of 1993, flood victims were compensated but were not permitted to rebuild in the floodplain. People with flood losses should get reimbursed only if they rebuild in a flood-free zone. Let's just say "no" to building on floodplains.

Endangered Ecosystems, Not Just Endangered Species

In order to address the importance and complexity of preserving biodiversity, we need to think about the entire ecosystem, not just one species at a time. An endangered species is only a tiny part of an entire ecosystem, and perfecting the requirements of only one species is shortsighted. For example, in the Longleaf Pine ecosystem, a critical niche requirement for the endangered Red-cockaded Woodpeckers is the older Longleaf Pine trees, which are isolated and have large cavities for nesting. The survival of these trees depends on fire and isolation from competing trees. Rather than concentrating on only the woodpecker, it is better to preserve the entire ecosystem, thereby protecting the entire suite of conditions that support Red-cockaded Woodpeckers, Longleaf Pines, and all the other species that interact in the community and support one another. This type of management should allow the natural and essential fire regime to continue as well.

The Endangered Species Act should be expanded into an Endangered Ecosystems Act. If an ecosystem is threatened by economic development projects such as road building or dam construction we must view the ecosystem as an intact whole before the work begins. If it is in an ecologically sensitive area, the project should be abandoned. We must seek to preserve entire systems. This is the approach now advocated by the Nature Conservancy.

Under the Endangered Species Act of 1973, we have been using endangered species as levers to force the preservation of entire ecosystems. For example, the campaign to save the Spotted Owl (*Strix occidentalis*) was an attempt to preserve the old-growth-forest ecosystem of the Pacific Northwest of the United States, where the owl nests in old-growth trees, notably Douglas Firs. Logging companies were cutting down the old growth firs, clearcutting vast tracts of them, thus eliminating the owl's nest sites. Environmental groups used the Endangered Species Act, and the threat to the Spotted Owl, to force a halt to such large-scale logging. But this species-by-species approach, while sometimes very effective and often gratifying, really misses the larger point. All species depend on their habitat and their ecosystem. Therefore, I suggest that the focus of international conservation efforts should be on the endangered ecosystems of the world.

Another example is the Giant Panda (*Ailuropoda melanoleuca*), which is threatened because of land development in its native China. There, bulldozers, clearing land for housing construction and agriculture, are devastating the great bamboo forests. Bamboo is the primary food of these gentle beasts, and the bamboo forest is their home. To protect them, we need to protect the bamboo forests they depend on. The Panda's very existence depends on the complex web of life that makes up the bamboo forest ecosystem: the growing bamboo and its plant competitors; the insects, bacteria, and fungi that decompose the dead leaves and stalks; the animals that help spread the seeds; the birds that keep the insects under control; the nutrient cycling; in fact, the entire ecosystem. The World Wildlife Fund uses the Giant Panda for its logo because people are more attracted to the large, extraordinary animals, the "charismatic mega-vertebrates," than the bamboo forest ecosystem of which they are only a part. We focus on saving the endangered Florida Panther, when what really needs our protection is the entire natural system, including the swamp forest ecosystems, the Everglades ecosystem, and the entire hydrological regime of South Florida.

Let's concentrate on saving ecosystems, not only individual species.

Living Eco-Logically: Individual Responsibility, Conservation, and Sustainable Development

As we consider the scope of environmental problems and the conundrum of sustainable development, it is easy to feel personally powerless. We ask, "What can I do to preserve species, protect biodiversity, and ensure the continued viability and health of ecosystems?" Sustainable development, human population control, ecosystem conservation, endangered species preservation—these are such grand ideas that they seem impossible to achieve. If we apply the principles of ecology to our daily lives, we are taking the small, but real, steps that will take us toward a healthy environment and a sustainable future. The more we learn about how our ecosystems, communities, and populations work, the more certain we will become of the next step to take. It starts with us.

Each of us must take personal responsibility for our decision-making and for the actions of our governments toward our environment and the natural world: how many children should we have and when; where to build and where to live; what to keep and what to throw away. As voters, we also determine the behavior of our governments toward the environment. Each individual, committing himself or herself to living ecologically, can make a difference. An ancient Chinese proverb says: "Every thousand-mile journey begins with a first step." Let's take that first step and live eco-logically.

Closing the Loop: The Linkage between Disturbance, Conservation, and Sustainable Development

We want the world that we leave to our children and grandchildren to be as healthy and beautiful as the world we live in, or even better. We want to raise the standard of living for the poorest members of society. To attain these goals, we must protect and preserve healthy, functioning ecosystems and their biodiversity. These goals must be linked, since to achieve a sustainable future, we must couple healthy ecosystems with healthy, fulfilled people.

Development is sustainable. Growth is not. We need sufficient land to support healthy ecosystems and preserve biodiversity. The size of the preserves must be large enough to support our most wide-ranging species and be linked with extensive natural corridors. This preservation includes maintaining large-enough tracts of natural landscapes so that disturbances can occur without impinging inordinately on human settlements.

As we accept the need for large natural areas, we must also accept natural disturbances as a part of life. We need the rejuvenating and restorative effects of disturbances: to create patchiness in the habitat, to permit the continued existence of biodiversity, and to maintain ecosystem health. Despite our best efforts, we cannot successfully control nature, and we must learn to live with it. We must understand and accept that ecosystems are dynamic, not static. Those fearsome fires, floods, and storms are essential to the maintenance of biodiversity and to the healthy functioning of ecosystems. We must develop our human communities with an understanding and appreciation of the critical role of disturbance in our ecosystems and our own life-support systems. We must reconnect to the ecosystem and live in accord with it. If we use our heads and accept nature's realities we will live eco-logically. Along that path lies a sustainable and ecologically viable future.

Further Reading

Kaufman, W., and O. H. Pilkey. 1983. *The Beaches Are Moving: The Drowning of America's Shoreline.* Durham, NC: Duke University Press.

Index